A Career Worth Engineering

Transitioning from student to professional can be a challenging journey, but it doesn't have to be a daunting mystery. In *A Career Worth Engineering: Don't Just Graduate—Navigate the Transition from Student to Profession*, the authors draw from their own experiences and share valuable lessons learned in the foundational years of their careers.

The authors offer practical steps and insights to empower readers to take control of their career paths. By sharing their experiences, they aim to help others navigate the challenges and uncertainties of transitioning from student to professional. This book is designed to accelerate career growth and help individuals achieve their full potential, goals, and confidence in the engineering field.

This book is a guide for new engineering students, recent graduates searching for their first job, or professionals feeling stagnant in their careers.

Continuous Improvement Series

Series Editors: Elizabeth A. Cudney and Tina Kanti Agustiady

Transforming Organizations
One Process at a Time
Kathryn A. LeRoy

Continuous Improvement, Probability, and Statistics
Using Creative Hands-On Techniques
William Hooper

Affordability
Integrating Value, Customer, and Cost for Continuous Improvement
Paul Walter Odomirok, Sr.

Design for Six Sigma
A Practical Approach through Innovation
Elizabeth A. Cudney and Tina Kanti Agustiady

Building a Sustainable Lean Culture
An Implementation Guide
Tina Kanti Agustiady and Elizabeth A. Cudney

Lean Sustainability
A Pathway to a Circular Economy
Elizabeth A. Cudney, Sandra L. Furterer, Chad M. Laux, Gaganpreet S. Hundal

A Career Worth Engineering
Don't Just Graduate—Navigate the Transition from Student to Professional
John S. Rogers and Sean Maciag

For more information about this series, please visit: https://www.crcpress.com/Continuous-Improvement-Series/book-series/CONIMPSER

A Career Worth Engineering

Don't Just Graduate—Navigate the Transition from Student to Professional

John S. Rogers and Sean Maciag

CRC Press
Taylor & Francis Group
Boca Raton London New York

CRC Press is an imprint of the
Taylor & Francis Group, an **informa** business

Designed cover image: Shutterstock – O-IAHI

First edition published 2025
by CRC Press
2385 NW Executive Center Drive, Suite 320, Boca Raton FL 33431

and by CRC Press
4 Park Square, Milton Park, Abingdon, Oxon, OX14 4RN

CRC Press is an imprint of Taylor & Francis Group, LLC

ISBN: 978-1-032-84034-5 (hbk)
ISBN: 978-1-032-84001-7 (pbk)
ISBN: 978-1-003-51090-1 (ebk)

DOI: 10.1201/9781003510901

Typeset in Times
by SPi Technologies India Pvt Ltd (Straive)

John S. Rogers: To my family, for always saying I could get here, and most of all to my wife, Monica, for helping me find myself again.

Sean Maciag: To my wife Sara for her endless support and belief in all my endeavors, and to my son Maksim for showing me what life is truly about.

Contents

Preface to "A Career Worth Engineering"

Sean and I were discussing our careers one day at work, and we made a mutual discovery: Virtually every lesson we had learned in our careers out of college, we had learned the hard way. Everything – landing jobs, getting promotions, changing industries, graduate school selection – everything had been learned the hard way. This realization sparked something in us as engineers and mentors, and we started to wonder if we could share our experiences to allow others to avoid some of the "hard way" in learning. Just like that, "A Career Worth Engineering" was born.

We set out to answer not only the questions that we had in the past when we were just starting out, but also the questions that came up so frequently from the young professionals we were now mentoring and developing in our careers and personal efforts. Questions on topics that almost invariably have come up, or that are very likely to come over the horizon of your future career too. We draw upon more than 15 years of combined experience as engineers, mentors, mentees, and learners, and we distill these lessons in a way that makes them approachable to you the reader, regardless of the exact path that led you where you are.

Along the way, we had ups and downs, plenty of people who were in our corners telling us we could make it, as well as a few people saying we wouldn't. We'd like to thank both groups for the motivation to get here.

Writing this book has been an amazing journey for us as authors, and we truly hope you find the inspiration to make your career what you want. Our goal was always to enable you to take charge of your career so that you can in turn lead and empower others. If there are two things we want you to take from reading this book it is this: You *CAN* make it through, and the best part of success is helping others.

Semper ad meliora – Always toward better things.

Foreword

I commend you for considering or embarking on an engineering career. I think it is noble and valuable work and I still love it. If you're made for it, there is nothing better.

Maybe you're here because you love gadgets or machinery, or maybe you're a math whiz and your high school counselor pointed you this way, or maybe you are a technician who is ready for the next step. There is a place for all of you to succeed in this field. You need to find the right fit of school, company, and job – and this book can help you with each of those.

There is such a variety of products, problems, and processes that engineers get the chance to invent or improve. You may save a life, enable people to thrive rather than merely surviving, or bring enjoyment to millions.

In my career so far I've been the new guy and the experienced mentor, I've been the team member and the manager, and I've been thrilled by my successes and humbled by failures. I've seen my designs on store shelves and I've had projects canceled right before they were to be released.

I've worked closely with hundreds of engineers of all disciplines (including one of the authors of this book) and mentored dozens. I've seen what works and what doesn't. I can tell you that the information in this book will help you in your education and career. It gives clear, actionable advice written in a fun, conversational style. It has information that may take you years to assemble on your own, and some information that might never get shared with you in the course of your work either because nobody thinks to share it, or your company would rather you not know it.

Your early career is important to set fundamental habits and expectations of quality work that should serve you for the rest of you career. These things are not taught in school because your professors may have never worked in industry and also every company or industry has different processes and expectations. I hope that you find a good mentor during your critical first few years of work. If not, this book and some creativity from you could create a good substitute.

> I hope that you find a fulfilling career and satisfaction in your life. Be curious, be persistent, and strive for perfection while knowing you cannot achieve it. This will bring you success.
>
> –Brian Ruff

Brian earned a BSME degree from Kansas State University in 1998. He has been a high-level individual contributor and manager in Mechanical Engineering for over 25 years at well-known global companies. He's designed military hardware, automotive components, aviation and consumer navigation devices, wearable electronics, professional loudspeakers, and laser projectors. He holds 17 patents so far.

About the Authors

John S. Rogers holds a Bachelor's degree in Engineering Management from the Missouri University of Science and Technology, as well as a graduate degree in Industrial and Systems Engineering from Auburn University. At the time of this writing, he is pursuing a "bonus" graduate degree in Aerospace Engineering from Mississippi State University. All things aerospace and professional development are his passions. You can find him on the weekends either cooking or planning travel out with his wife Monica, or you may just bump into him 65 feet below the water diving.

Sean Maciag is a first-generation college graduate and holds a Bachelor's degree in Mechanical Engineering from Arizona State University. In addition to Engineering, Sean is passionate about Brazilian Jiu Jitsu and imparting his knowledge of the sport onto others. You can find him alongside his wife taking their son to the park or traveling back to her childhood home in Eastern Europe.

Writing this book has been for us, an extreme joy and a ton of fun. This project came to fruition one day when Sean and I were just talking about things we liked, and among them, how we wanted to expand our mentorship audience. We realized there was nearly no resource we could point early career folks to universally, and certainly no resource that had the detail we wanted to produce on so broad a range. Our goal in writing this was to compile the resource we wished we had when we were in your shoes and to relay the lessons we learned in a manner that would, hopefully, save you some of the pain and headaches we got from banging our own heads against the walls we have run into.

1 Introduction

How could this have happened?

Sure, I hadn't been the best of students over the last year, at least not according to my grades, but this shouldn't have happened. I was a straight-A student in high school, and my dad was an engineer. I was supposed to be a second generation engineer, so how could I have possibly found myself here now staring at the piece of paper that said I had been removed from my undergraduate program? I was in my junior year by hours, which meant I had to have a declared program. Getting kicked out of aerospace engineering at this point was tantamount to just being kicked out of school altogether. A short two days after my 22nd birthday, and only a week before the start of the spring semester, I found myself a student without a school. Whatever led to it, it had happened. I needed to act, and act fast.

Engineering is not only my career, it is the source of some of my highest highs, and of my lowest lows, as evidenced in the brief story above. We are sure some of you have a similar experience, and if not to that level of panic, surely, you've faced some problems you weren't sure how to move past. Problems that made you ask yourself "How did I get myself into this?" The good news in this is that you are certainly not alone, and if you're in a situation that causes you some level of panic, we can help you chart a path forward. Despite how it feels, failures don't define you, and if you approach them correctly, they can provide growth like no success can.

As for how I got myself into (and more importantly how I got myself out of) the situation I was in, we'll address that later on in the book. We open with this story to let you know a few things up front: (1) Engineering is a phenomenal career, but getting there is more marathon than sprint. (2) It's okay to get stuck and not know how to move forward from time to time, but the important thing is you don't stay stuck. And (3) Growing as a person is rarely comfortable, but it is always beneficial.

We authors were friends before we became writers, and coworkers earlier still. This book is the result of two engineers who decided to be the resource we needed when we were just starting out. If you've found yourself facing a problem related to your career or education, we've been there too, and it's something you *can* get through. The intent of this book is to give you practical guidance through some common areas you'll progress through between engineering school and the first five years of your career or so, drawing on our own experiences and occasionally, failures.

Let's back up for a minute. What if you haven't dug yourself a hole, and maybe you're not even sure what an engineer is? Maybe you're still in your high school career and are n't sure what you want to do at this point, but you keep getting told by your teachers (or maybe your friends and family) that you should be an engineer. You may not even know what engineering is, but because numbers and math come easily

DOI: 10.1201/9781003510901-1

1

to you, or at least because you're interested in making things and finding out how things work, you've become notionally interested in engineering. Like so many others though, maybe you aren't quite sure where to go from here, and certainly not sure how to make a career out of engineering.

Possibly, you're coming into this book even earlier than that, and you're unsure of what you want to do with your career, and that's also okay. We intend to share with you not only what engineering is, but how you can use it to lead your career where you want it to take you. Even if you're not an engineer or planning on becoming one, stick with us! The actions and methods we present in this book will still apply to all kinds of professionals leaving college and entering the workforce. Whether you are a fresh out high school graduate with no clue of what to expect in the professional world, or you're three years into your career and feeling stuck in the mud and spinning your wheels at your current employer – every step of the way, this book has something for you.

Engineers at their core are creators and they are problem solvers. We take what does not exist and we bring it to reality – something fun to think about is that nearly everything you see around you right now has been engineered, by someone, somewhere. From the computers we typed this book on, to the couch you may be sitting on, to the espresso machine that made the coffee you're drinking, nearly everything started life as an idea that was put on paper (even figurative paper!) then carefully drawn and created. A problem was presented, and engineers provided a solution. The core of engineering is that problems need solutions, and that core is fundamentally beautiful in its simplicity. Before we get to what engineering is and why it's such a great field to get into in much more detail later, let's talk about who we are and our path that led us to engineering.

I (John) came into engineering in a more straightforward way than most. Having grown up the son of an electrical engineer, I had the one-two punch of being not just an engineer's kid, but also of growing up in the Cradle of Naval Aviation, Pensacola Florida – home of the US Navy's Blue Angels flight demonstration team. As a child, I was constantly bombarded by the sights and sounds of powered flight, fostering my love of any and all things aviation. From there, I ended up doing my undergrad in Engineering Management then onto a Master's in Industrial and Systems Engineering (War Eagle!), but not before that brief stint as a student without a school. Ultimately, I don't know how far I'll go in formal education, but I would gander a guess and say I have more stops to make before I hang up my calculator. At the time of this writing, I am underway on my second Master's degree, this one in Aerospace Engineering. My entire career to this point has been in project and program management, over a broad range of products and industries. I am in the aerospace industry, which I have known I wanted to be in since I was a child. Though I may have been exposed to engineering earlier, my journey into it was largely no different than yours may well be – overcoming your lows, learning how to navigate the transitions that come from leaving school for the workforce, pursuing higher education, and learning how to earn promotions or change jobs are all skills that we learned, and we now intend to share with you.

When I (Sean) was young the question most asked seemed to be "What do you want to be when you grow up?" The younger I was, the easier it was to answer.

Doctor, lawyer, engineer, firefighter, police officer, astronaut, pirate, ice cream taster, the list goes on! It was sometime in my early teens that I realized this wasn't a question of dreams and imagination, but rather a call to really look inward and tell myself "I want this career to define my professional life for the next 40+ years." That was a serious pill to swallow at fifteen!

Luckily, my high school provided us with a career aptitude test and "engineer" was my #1 match. I wasn't quite sure what exactly that entailed, but with a bit of research I decided, "Hey, I like knowing how things work and am good enough at math, I can probably pass these classes and earn a Mechanical Engineering degree." In hindsight, I was lucky that my very loose criteria placed me on a path to one of the most fulfilling careers I could possibly imagine, providing real value for myself, my company, and the industries in which I have worked. While everyone finds their career at a different pace, whether it be at 15 or 40, we must all walk a similar path to attaining the necessary education and following the process from student to early career to established professional.

Now, the story of how you found yourself headed on the path of becoming an engineer will most definitely differ from John's and my own, but you are along for the ride all the same – highs and lows, triumphs and frustrations. We bring 15-plus years of combined lessons learned and share them so you can triumph too.

Enough about us, let's get you back to the real reason you're here: Engineering. We have briefly touched on what engineers do, but if you're like a lot of people who didn't grow up around engineering, you may be confused about what they truly do and what their jobs entail. I know that I (John) was a little surprised when my wife told me I was the first engineer she had met once we got together. What gives? Well, engineers aren't in everyone's circle of acquaintances, and sometimes they can be as shadows in the background – working but not in the forefront of people's minds. Nevertheless, the world has problems, and it is the engineer's calling to solve them.

2 What Is Engineering

First of all, there are an abundance of engineering disciplines (covering everything from textiles to radio waves to lasers to spacecraft to … the list goes on to virtually anything you can imagine), but generally speaking, there are four major categories of engineering: Chemical, Mechanical, Civil, and Electrical. As stated, there are a lot more, so if you can dream it, there's probably a program for you, but these four are a great baseline to get started with. Broadly speaking, Chemical Engineers make processes, Civil Engineers make roads, buildings, and bridges, Electrical Engineers make power and electronics, and Mechanical Engineers design nearly everything else (no really, they do). Beyond that, there are specific specializations, such as a Civil Engineer becoming a Structural Engineer (loosely akin to thinking how all orthodontists are dentists, but not all dentists are orthodontists), and further specialties across a range of specific disciplines, like Aerospace Engineering, Systems Engineering, Hydraulic Engineering, Tribology, Metrology, etc. The sky is truly the limit to not only the kind of engineering work you can do, but also as to *what* you can do with an engineering degree – we know medical doctors, attorneys, authors, and even small business owners who started out as engineers first. That, to us, is one of the single most under-realized aspects of an engineering degree: The ability to solve problems given the resources and knowledge that you have at your disposal. This problem solving is a central theme that we'll revisit throughout this book, and we encourage you to consider in your own life, as viewed through your own lens.

Let's look at a representative case for engineers. If someone asked you to design an aircraft, where would you start? What about a car? What about a replacement heart? Engineers tackle a broad range of challenges, but they do so with the building blocks they started learning in school, so they don't have to reinvent the wheel every time. Engineers take what they have learned and apply it to problems as they appear. In fact, that is one of the main differences in a question you may have been wondering: What is the difference between an engineer and a scientist? Engineers are more focused on "knowing what we know, how can we make this work?," while pure science is more focused on answering the question of "what do we know?."

Once you know what they are, the next question you may be wondering is what do you have to do to get there and join their ranks? Well, while there are pathways for someone to become an engineer outside of the classroom, the most straightforward and readily available path is found at any number of accredited universities. Within these institutions you can pick between any of the aforementioned engineering disciplines and customize your education through something called "technical electives," which allow students to fine tune what they are really searching for in terms of employment after graduating.

DOI: 10.1201/9781003510901-2

What might a career progression in engineering look like? Once you've completed your education, you'll encounter that professionally, engineers who stay on the technical path – that is, those who do not go into some form of management – follow a generally defined hierarchy of seniority, which will loosely align with the following:

- Associate Engineer: 0–2 years of experience
- Engineer: 2–5 years of experience
- Senior Engineer: 5–7 years of experience
- Staff (or Principal) Engineer: 7–12 years of experience
- Senior Staff (Principal) Engineer: 12+ years' experience
- Fellow: typically in the realm of 25+ years of experience, dependent upon knowledge/contributions to the field. Generally, fellows are very rare, and some organizations may not even have this position.

The above isn't intended to be a complete list, but rather to give you some perspective of how companies consider engineering experience and titles, with the primary focus of this book being students in either high school or college, up to about five years of professional, post-college experience, as we stated earlier. We also note that the titles you see may differ, depending on industry. For example, a Senior Engineer may be a Lead Engineer, or all the titles could be boiled down to Engineer I–IV, but the separation between years of experience and recognized level of competency will mostly remain the same. As with most things in Engineering, there is more than one way to approach your career and years of experience can often be substituted with additional degrees (usually, with a master's degree equating to two years of experience, while a doctorate can land you in five-years of experience territory. This largely depends on the industry and what you were doing in terms of research and application in such a doctoral program) in terms of pay scale and expected responsibility. There is also the availability to move into management of other engineers or programs, which will of course have their own paths.

From this point, there are nearly limitless options, and as time goes on, you'll find yourself asking "what do I do now?" and wondering if you should move from one company to another, change from design engineering to testing engineering, or even into management, or perhaps what promotions to pursue. While we would love to tell you what the perfect route is to take in order to maximize your success and satisfaction, the truth is this is your career and finding your calling is half the battle. Don't worry though, we are here to help you with the "how" to get you there, based on the lessons we've learned the hard way. Again, our intent with this book is to provide a resource that we wish we had walking into our freshman year of Engineering School. From the opportunities you have while in university to finding that first job to transitioning through the first major opportunities in your engineering career, we have been there and can provide insight to guide you.

Now that engineering is kind of loosely defined and you generally know what it is, what happens next? Well, if you choose to pursue it, what happens next is a lot of schooling – a whole lot of schooling, actually. Now is a great time to mention a fact about engineering school: It will be hard. Worth it to most people, but hard no less. One of the single most important factors we can think of regarding someone making it through engineering school is simply them wanting to do it. Neither your parents, nor your friends, nor your teachers can want you to be an engineer enough to make you get through the time, effort, frustration, and personal determination it takes to earn an engineering degree. Only you can make that decision, and sometimes you'll really question it, but for so many people, the moment you make it through, the reward and joy are like no other. We know our feeling of accomplishment when we finally saw "degree conferred" across our transcripts was like no other feeling of elation. Doing hard things often yields a reward all its own when they are done.

That brings us to an excellent opportunity to introduce something that in our minds really is not addressed all that often: You can be an engineer if math is not your favorite subject, or even if you're not naturally "good" at math (what does that even mean, anyway?). There will be math, there will be hard math, and you will have to get through it one way or another, yes. Those are all facts. I (John) struggled a lot in calculus and differential equations early on and I am an outright terrible computer programmer. Yet, even still, I made it through a pair of engineering degrees that required both math and programming. That is also a fact. We'll hit on this topic more later, but for now, we want to put it on paper that you are not immediately disqualified from being an engineer because you don't see the world in hyperbolic trig functions or in pure terms of undefined variables. If you have a good understanding of mathematic principles, and you're willing to put forth the effort you need to succeed, you'll learn that you can dig deep and make it happen. Another advantage you have now is the level of resources available to you as students seems to grow every year. Everything from internet resources to social media is available to help you understand concepts. If you're not following it the way your instructor is explaining it, there's probably someone else out there explaining it another way. We mentioned that one of the greatest attributes of an engineering degree is gaining the ability to solve problems with the resources you have at your disposal. Engineering school is mostly a mental game in more ways than just one.

Once you have decided to pursue engineering, the academics are certainly tough, as we mentioned, but there is a silver lining to this pain and suffering. In addition to your own satisfaction and the reward of succeeding at something difficult, engineers have two main career draws: Jobs and (typically) good salaries. There are a lot of engineers needed, and good ones can earn great salaries over their careers. This, coupled with the opportunities to work on projects that range from the top secret, bleeding edge technologies all the way to the projects that impact the quality of life for millions of people at scale means that engineering is never short of opportunities for you to find your calling.

Ok, now that you're ready to chase after that engineering degree, you need to do some schooling, and you may have some questions regarding how to pick a school. Here we have some general suggestions: Core classes (math, language, humanities, and the other classes that make up your first two years of college) are going to be essentially the same (more or less) whether you are in the most prestigious Ivy League school, or your local junior or community college. We say more or less here because in our experience, if anything, these classes are easier at smaller academic institutions, partly because of class sizes and student-to-instructor ratios.

If the expense of studying for school is a concern, you can save yourself a lot of money by considering the two-year institution route and then transferring to a four-year institution to finish the in-major portion of your degree. In fact, depending on your state and some contributing factors – high school grades, standardized test scores, etc. – a junior college may even be free or close to free for you to attend. Getting half of your degree out of the way for free or nearly free is a huge benefit later on if you still end up needing to pay for your final two years. You may be thinking "but I really want to go to Rah-Rah University for the full experience," and you can, but the truth is calculus is the same in every school and in every country (and has been since the 17th century, when Newton and Leibniz first started describing it). The reality is, the most important thing you can do for yourself regardless of where you start your engineering degree is getting a rock-solid foundation for what comes next, or you will have a much harder time getting to the finish line than you would have had otherwise.

One thing to keep in perspective during this is that college will always have a cost associated with it, and that cost can vary hugely from school to school. We are not at all advocating you don't go to highly regarded schools, or even the school you want, but rather issuing some caution about the cost-versus-benefit for your given program or school. The good news is that fairly easily, you can see not only the expected earnings and growth potential of a given degree with a US Government agency – the Bureau of Labor Statistics[1] – you can also check most school's expected earnings for a program's new graduates, mid-, and late-career folks through various publications like US News and World Report among others (websites in the endnotes). This information can be extremely helpful in helping you decide on where you go to school and what program you choose.

Another word of caution we give you: Please be cognizant of the loans you take out, if you must do so. We certainly understand that school is a huge endeavor and not everyone is so fortunate as to have it paid for by someone else. Student debt is a fixture of higher education that will likely be here to stay until society figures out a better way to deal with it, so we recognize loans can be a useful and necessary means to an end. There exists plenty of excellent outside literature surrounding this topic, so we'll leave it fairly simply: Be absolutely certain you understand the terms of the promissory note you sign to pay it back, and *absolutely* do not take out more than you need. Luxury goods can wait until after school. Part of the reason behind our writing this book is to help all engineers succeed, and while success is multifaceted, there is no sense in starting with more of an uphill trudge than you need to have.

It may be far off in the future, but if and when you start considering graduate degrees in engineering, my (John's) advice is to find the biggest name university you can for what you're wanting to pursue and go from there. Reason being? Bigger schools have (generally) bigger budgets for research, which is what drives a lot of graduate engineering programs. More on that later. PS, it's never too early to start thinking of your future plans. I (John) knew a graduate engineering degree was in my future as early as about 15 years old. I am now one graduate degree in, and have started to pursue my second graduate degree at the time of this writing, which is to say, sometimes you just know you're going to keep going and that should figure into your plan.

Something that is not discussed very often is the overall topic of *how to pick a school*. Sure, engineering is great, and you've picked a degree program that you want to pursue, maybe even narrowed it down to a few finalist universities to pick between. How do you pick? It's going to be a combination of factors, and we wouldn't say any of them are necessarily the final word, so it's going to depend on you. Having said that, there are certainly some things that would make a school standout: How is their overall ranking and accreditation for the programs you're interested in? What kind of costs/financial aid packages are they offering? What noticeable things did you connect with the faculty/facilities/campus? We go into more detail on these types of things in the section titled "Things we wish we knew," later in the book. Make sure though, that the institution you pick is a good fit for you, and not just a good fit for what someone else wants you to do, and also consider the total cost of attendance. Good financial aid packages may lose their luster when compared as a whole picture.

The road to an engineering degree is a long one already, so doing a program that your heart really isn't in will make that road seem nigh impossible. It's also a good idea to align with what the school's core competencies and focal areas of research when and where you can, though we fully understand that is not always feasible. What we mean here is if you wanted to study aerospace engineering, for example, we can point you to a list of schools who have an aerospace program which is much more related to naval architecture and shipbuilding than it is hypersonic atmospheric flight. Because it's fluid mechanics either way, it technically still makes sense but may not be what you are looking for specifically. Furthermore, not all aerospace schools have a lot going on with, say, helicopters or VTOL (Vertical Take Off and Landing) flight, so if that's what you're set on studying, you may need to narrow your search. It's okay if finding a school takes some time, because it's a big decision which will guide a lot of your decisions over your next few years. Make sure to do your research on this and discover what the institution does, and know this may take some effort beyond just looking at their website on your part.

Over the course of the coming chapters, we will guide you through what the process looks like from engineering school to about five years after graduation, including steps like how to make the most of your college experience and craft a good resume to help you land that first job. Through all the trials, struggles, doubts, sleepless nights, and questioning, on the other side we can still say we absolutely loved our journeys to where we are today, and we're here to help you as you embark on your own journey. We truly hope you benefit from this book as much as we have benefited in writing it.

CHAPTER SUMMARY

If you can dream it, there's probably an engineering discipline that addresses it. Engineering can be a great career, but it's going to require some work from you, and you're going to have to want to stick with it. If you do, there is a whole career just waiting for you to seize and make your own.

- Engineers make nearly everything
- Growing is rarely comfortable
- It's okay to not know how to proceed and need some help from others
- You can still be an engineer even if math doesn't come naturally to you

NOTE

1 https://www.bls.gov/ooh/field-of-degree/engineering/engineering-field-of-degree.htm

3 Opportunities in University

Now that you have – through a combination of hard work, decision making, term papers, and more than your fair share of suffering through things you didn't feel like you should be spending time learning – finally gotten into the specific engineering program and school of your choice, what comes next? First of all, take a second to breathe and congratulate yourself. You made a goal, started working toward it, and are now seeing some of the progress pay off. This may seem like a small win but take it from someone who spent way too long not celebrating anything that it's good to reflect and be proud of what you've accomplished so far. The road to engineering school is a long one already, and for some it is harder than others. I (John) mean it sincerely. Take some time, reflect on getting this far, but know that the hard work is just starting.

Engineering, like so many other things in life, is going to largely give you back what you put into it, *especially* in the early days as you are starting from scratch and learning your new craft. As part of that, it is absolutely crucial that you happen to college, and you don't let college happen to you. What we mean there is if you are not going in with a plan, it can be easy to slip into a routine and not get the most out of your time while enrolled.

Academically, you may already have an idea of what to expect as you pursue an engineering education, or you may not. We're going to cover the *general* flow you may encounter, and it's honestly not all that different than what you may have seen with other programs or degrees. You'll typically have a total of somewhere around 130 credit hours you'll need to take to graduate, and among those it works out to roughly half general education and half what is in your major. You'll notice that general education for engineers is different than general education for other degree programs, as they typically have less stringent math requirements, or at a minimum, don't typically take three semesters (or more) of calculus and calculus-based classes. Typically, all engineering programs require three semesters of calculus, plus differential equations, and then most require two or three semesters of calculus-based physics and other classes. This means generally speaking, you need to have completed things like trigonometry and college algebra prior to starting your engineering curriculum, or you'll need to get that squared away pretty quickly once you transfer into a program. Another thing you may have heard of to this point is college degrees being referred to as "four-year" degrees, or that universities are "four-year institutions," because it's kind of the traditional expectation that college takes a person four years to finish. Don't get too caught up in this or feel like you're a failure because it took you more than four years to finish. Engineering is a lot of work, and if you take co-ops or internships, you're likely going to push that out a semester or more, and if you need to retake a class, you can add some time to your college trip. That's okay too. The four-year college concept

DOI: 10.1201/9781003510901-3

is typical across the board for everyone, including for non-engineers, and typical is not what everyone is all the time. If you do find yourself taking a class over, chalk it up to a learning experience and move on. A lot of us have retaken classes, including me (John). The important thing is that you continue to press on and don't let yourself get discouraged, because engineering school can be a long road – one that is definitely a marathon and not a sprint.

If you're not fully decided on a program yet, the good news for you is that the pre-requisites for most programs are quite similar; even if you didn't plan on going into engineering originally, you may not even need to add that many classes in. For example, every college student will take some humanities classes and composition and literature classes as well, even engineers. The most important thing you can do for yourself is to have a good foundation in the "core" of math and science, because that will be a recurring theme for your entire education. Your path will then involve around two years of "gen-eds" (general education, because *generally* everyone takes them), then around two years of what's often called "in major" classes, and to us, as fun as learning is, these are where the *real* fun begins. This is why you decided to go into engineering school in the first place, and here you finally get to do engineering. Here, depending on your major, you'll get to take relevant classes which so often build on each other. We encourage you to attend classes and do homework/practice problems as much as necessary for you to understand the material, even if it's not required. Because so many concepts in engineering are built on prior understanding, youll see the same things over and over, and it's in your best interest to get those locked down as early as you can. You don't want to do what I (John) did and not get a grasp on statics until you are in mechanics of materials (the class that comes next). Trust me, as a person who regretted not doing more homework when I could, there are no shortcuts in engineering school. As one of my instructors (for statics no less) said, "get the points while they are easy." We can guarantee that your upperclassman self will appreciate and be grateful for the effort you put in these foundational classes during your freshman and sophomore years.

In any case, your path through engineering school will likely be one of the most challenging things you've done in your life to that point, and unless you're just one of those people who "gets it" from the start, you're going to question yourself more than once. That's okay, and we want to stress it is totally normal to not be one of those "just gets it" people and to have some struggles and doubts as you work through it – maybe even a lot of them. We want you to have a firm foundation to start with, so the main thing is that you put forth the effort and time that you need to devote to the education, and not get too caught up in seeing college as only a fun time. It is a means to an end, and you can absolutely enjoy your time there, but you're there with a goal of leaving at some point and taking all that you've learned with you out into your career as an engineer.

Because a full education is so much more than just academics, there are a lot of things to keep in mind as you start working on the next steps. One such thing to keep in mind is that colleges and universities are excellent places to start developing your skills, both academically and outside the classroom. Depending on the program or major you have chosen, as well as the specifics of your particular institution, you likely have a lot of opportunities available to you that you may not even realize. Part of what we want to teach you are skills you can use to identify and process opportunities to open doors later on, and you can begin honing those skills in college.

To that end, student design teams are a wonderful way to connect with others with similar interests, build your skills, and just as importantly, meet professionals in that arena. In my (John's) experience, even smaller junior and community colleges have some level of engineering clubs, societies, and teams. If you're not sure where to start, begin with your faculty or the school's helpdesk and go from there. They likely won't have everything, but a great mentality to have as you begin your engineering career is that any experience is good experience. My junior college had an engineering society, and while we didn't do anything earth-shattering, I was fortunate enough to listen to some pretty interesting guest lectures, meet some people, and tour facilities like the John C. Stennis Space Center in Mississippi and the National High Magnetic Field Laboratory in Florida, both of which were really interesting places to see as a bright-eyed 18-year-old. I also dragged my parents to the National Naval Aviation Museum in my hometown every single chance I could get, and that really shaped my deep love of all things that fly. Clubs and teams also have the added benefit of giving you a larger acquaintance base to draw help from when you need it, as well as gain exposure to a broader field of engineering. If you're on a campus with truly nothing available, nothing says "leadership" quite like leadership, so start a club. Be creative, and don't be shy to reach out to faculty, alumni, and even local businesses, to help you get started. Spinning things up from a dead start is a lot of work, but it can be a lot of fun, not to mention a great place to sharpen your skills.

While maximizing your involvement with additional resources is important, it is imperative that you devote full effort to your classes first and foremost. I (John) was one of those classic "smart kids" in high school: I swam competitively since before high school, I had a 4.0 through graduation, and I had a pretty decent American College Testing (ACT) score. I NEVER put forth full effort in high school because I quite honestly didn't need to. I never even put forth all that much effort into my junior college classes because I didn't need to then either. Then, after earning my AA, *real engineering school* began, and let me be the first to tell you, my GPA got absolutely *obliterated* for two semesters. My junior college physics professor had this policy where you basically would get no lower than a B as long as you tried a problem. His definition of "tried" was pretty loose, and twenty-ish year old me did not have an adequate understanding of how much of a disservice that was to me. Free B without trying? Great, sign me up. In doing so, I dug a hole that took a whole lot of work for me to get out of, and I didn't even know I had done it at the time. At least, not until I got kicked out of the department as I mentioned in the opener of the book. More on that in our "Fighting the Impostor" chapter – it's quite a story.

For the present, let's get back to how you can maximize your time in school through clubs and societies. Engineering is so much more than just academics, and if you haven't gathered to this point, it's hard. Really hard. If you're going to make it through, you're going to need some people in your corner, and as I said, clubs/teams/societies can be an excellent way to get people in your corner. You can do things that are very "in the wheelhouse," or you can do things that may not line up so simply. What do we mean here? Let's say you're a mechanical engineering major, and you're really interested in race cars. Pretty logical right? You'd just join one of the SAE

(Society of Automotive Engineers) teams your campus has, with the two biggest being Formula and Baja (a scaled open-wheel style racecar, and a purpose-built offroad vehicle, respectively. I (John) was a two-year member of the former during my undergrad). A "linear" (or presumptively logical) club for someone with a civil engineering major could be a team like Steel Bridge, or Concrete Canoe, while an aerospace engineering major may go to something like a model rocketry or a satellite building team (I [John] was also a two-year member of a satellite building team during my undergrad). A computer or electrical engineering major may be interested in something domain-related for themselves.

Easy-peasy.

What if you don't fit as neatly in a box? Suppose you are a computer science or computer engineering major but you love racecars? What if you're a civil engineer and you love airplanes? What if you're another engineer and you don't know what you like? Fret not! Now is a great time to mention how interconnected engineering is. During my time on Formula SAE, we had Comp Sci's and Comp E's doing all kinds of simulation, data acquisition, and computer modeling for us. When I was on the satellite team, we had environmental engineers doing analysis, and both teams had civil engineers doing structural analysis. We had chemical engineers laying up carbon fiber for us, and we had mechanical engineers welding. We also had history majors welding. **Our most important resource and valuable currency was having dedicated people willing to show up and learn**.

See a pattern here?

It's exactly the same in the real world. I (John) have civil engineer friends who work in the aerospace industry, and while I have very limited software experience, I still lead teams and projects with almost only software people on them. One of the points we truly hope you take away from this book is that **engineering is so much more than a degree**. Engineering is proof you can follow a logical path and solve a problem, and that by itself is so much more valuable than just a degree. I have met plenty of engineers who later went down other paths, from medical doctors to restaurant owners, and they are all united by their ability to solve problems. A lot of them really refined their early problem-solving skills on design teams. That's the beauty of them: Problem solving is the beauty of engineering.

School is a time to broaden your horizons, and that includes not just socially or academically. Try things. Learn things. Try a team, find a new club, go do something you don't know if you'll like. Whatever you do, make sure you actually get your hands dirty. The first time I ever saw a lathe in real life, I was shown how to use it and given some calipers and a print, and told to start making parts for our Formula car. If you want to learn real skills that real employers want to see, you've got to get your hands on more than just books and paper. Design teams are a place for you to learn before you're getting paid for things. Once money comes into play, the dynamic changes. I've had great employers in this regard, but I know that isn't always the case. If you want to make an omelet, as the saying goes, you're going to have to break some eggs. It's better to break those eggs early and often in an environment where that is expected and encouraged. You can learn anything in life, as I truly believe the sky's the limit, as long as you're willing to put in the time and effort – as well as know that on occasion you'll break some eggs.

This leads into another excellent and often under-utilized benefit of design teams and extra-curriculars: The social aspect of them. Sure, humans aren't made to be alone and you're going to want (and need) friends during your journey in school, but in engineering clubs/societies, you'll also find yourself surrounded by those who are now your peers. Engineering peers serve a multitude of purposes. In a school setting, some of your peers have already learned what you're in the process of learning, some are learning with you, and others will be the ones following you. Learn from them. Learn with them. Lead them. For those ahead of you, ask all the questions you can about classes and instructors in your major, learn about electives and things you can learn as periphery to your core studies. For those who are learning alongside you, do homework assignments together to the extent allowed by your school and specific instructors (we are absolutely NOT telling you to cheat or otherwise violate your school's honor code. Cheating is a disservice to yourself, and it is not worth it). Explain concepts to each other. As others come up behind you, teach them what you have learned, like shortcuts and rules of thumb, plus ballpark estimates and such. The reason here is that our peers help us grow. Sometimes we learn from them, sometimes we teach them. In the end, everyone grows together. I (John) had a manager who loved the phrase "a rising tide lifts all boats." The same will be true as you progress beyond school to the workforce. Your peers are going to be a major source of learning, as well as how you can gauge where your own skills lie. We absolutely cannot overstate how important it will be to you to start developing your peer network while in school, and an easy way to start this is with student design teams.

Another extremely valuable aspect of design teams/clubs/societies is that they tend to attract industry professionals, many of whom are themselves alumni of the same clubs. One of my first team meetings with the Formula team was actually a guest speaker from one of the most prominent aerospace companies in the world. One of the members of the team had interned for them for three summers at that point and went to work for them immediately upon graduation. At the time of this writing, he is still there over ten years later. Certainly, this is going to be more easily done for teams and schools that are more established, but even if your team is in its infancy, you'll likely be surprised how many seasoned professionals are willing to lend a hand and help get the next generation of engineers up and running. The additional exposure you can get within a given industry is extreme, and this doesn't even count the students and faculty that you can meet at events and competitions.

Speaking for Formula SAE, the competitions were phenomenal places to learn and grow. The judges were industry professionals, and the way the competition worked forced you to truly know what was going on. Certainly, a healthy part of the competition was in the actual performance of the car, where our work would speak for itself, but a lot of points were to be made in the presentation portion of the competition. When an industry professional asks you why you run the brake fluid you chose to use, for example, you're going to need to have a better answer than just "it says brake fluid." I use this as an example because I have actually heard this question asked. (The real answer, by the way, is without doubt one of the funniest stories I ever heard during my undergrad, but that's for another time.)

Needing a real answer is true for other aspects of the event, for example, I gave a presentation with the student I mentioned earlier, the one at the major aerospace company. He was our outgoing business manager, and I was the incoming one, having been voted in by the team. This experience gave me a lot of chances to learn the "less fun" part of engineering, namely the business behind the scenes that makes the numbers work, and that is a huge part of getting promoted later in your career – doing the "less fun" work. Anyway, back to this competition. He and I came up with this presentation, which we delivered in front of a four-person panel, among them executives for two MAJOR automotive companies (ever heard of a pickup truck named after a dinosaur?) We would absolutely never have gotten this kind of opportunity in the classroom to present to major industry executives. What was our end result? We took 4th out of more than 100 teams at this event, which was both the highest scoring performance in school history, and the best presentation given by engineers at the competition (the others who beat us were teams comprised mostly business students with very elaborate presentations. They certainly deserved the win, but we were thrilled with our score). As I said before, if you are early in your career, any experience is probably going to be good experience.

Experience, as you'll find out, creates opportunity. Presumably, if you're going to school for engineering, you're going to want a job in engineering at some point. I can honestly say my design team experience is what got me my first Co-Op (a Co-Op is a kind of extended internship, which typically runs for a summer and a school semester, while the latter usually runs just a summer), hands down. I know this because my manager told me that directly. He knew from my experience that I was willing to get my hands dirty with work and that it wasn't just paper knowledge. That Co-Op got me an internship at a different company the next year, and both combined to land me my first full-time job after graduation. The bad news for landing jobs (even internships) early in your career is that you'll find out GPA plays a disproportionate part compared to later on in your career. Because you have less overall experience, potential employers have less to judge your potential on, so grades have a lot of impact. I personally don't think GPA is a perfectly accurate measure of performance, but early on and without more information, it's a necessary evil of sorts. Is your GPA less than stellar? That alone isn't an automatic reason you won't get any internships or co-ops necessarily, but it does mean you're going to really have to sell other things, like your extracurriculars and experience on teams and clubs. I (John) had a less than stellar GPA for a lot of my undergraduate degree, and I still landed internships, a co-op, and a full-time job after graduation. It absolutely can be done. More on that to come in our chapter on Chasing the Impostor.

We want to mention here that GPA is not a reflection of you as a person. GPA is a reflection of how well you can demonstrate mastery of a subject, which in and of itself is a subjective thing. Maybe you don't do well on tests, maybe you need to put more effort in, or maybe you need someone to explain it differently to you. All of these things are valid, and while GPAs will never tell the whole story of you, they are an important part of your college experience, so we want to make sure you're equipped with the all the resources we can provide you, as well as tips and strategies you can use for success. We'll cover this in some later chapters.

What's the upside? GPA aside, pursuing internships and co-ops while in school does a few things for you. First, it gives you the experience you need. I don't mean this as simply as gaining job experience, because to even get there, you're going to need to exercise some skills that will be important later on. For example, you'll need to learn how to interview and advocate for yourself. You may need to take advantage of services your school offers, such as getting resume help, attending career fairs, or reporting co-ops. All of these are skills, and it is better to learn them sooner rather than later. Again, any and all experience at this point is good experience. Another upside to pursuing employment while in school is it can open previously closed avenues to you. You may land a job at a company that loves to hire former interns, or you may potentially get exposure to a more specialized set of skills either in your degree or parallel to another, and be able to really focus your effort moving forward.

We want to also set some reasonable expectations for you as you read through the book: One being that comparison is the thief of joy. If your friend landed a co-op during their freshman year, that doesn't mean you are a bad student, or never going to land one. In fact, co-ops and internships as freshmen are somewhat rare, so don't let that be a point of contention. Rare doesn't mean impossible, so if they did manage to land one, learn what led to that. Were they really good at something, did they have a unique skill set, or did they just legitimately luck into the right place at the right time? Generally speaking, you're going to need some level of engineering course work to land an engineering internship, so while I fully encourage you to pursue them as soon as you can, do so with the understanding that your experience has to match the requirements of the job. If you don't land one right away, you still gain benefit from pursuing them. You may learn you really need to take a class to work at a certain company, or you can start networking with the companies you think you're interested in. Another expectation you should have is that nobody is on the same timeline. Whether you're a traditional student or a coming back to school later on, we're thrilled that you've made the choice to join the ranks of engineers. I've (John) known plenty of friends, colleagues, and coworkers who have come to school later on, or even earned a degree and gone back to pursue an engineering degree as a second one, and all of them have made meaningful contributions. Again, comparison is the thief of joy, and just because your timeline is different, there is no reason you can't succeed on this journey.

Something I (John) noticed throughout my undergraduate career was that a lot of the attendees of career fairs were the same people year after year, so you can really get to know them and learn about the company. I spoke with one company for nearly 3 years before I got an internship there, in my senior year. (I also didn't get a co-op until my junior year so don't worry too much if you're still looking!) Learn what you can, take notes on what you can improve upon, and make an effort to accomplish your goals, but absolutely keep things in perspective, and remember that you aren't a failure if you don't work every single summer while you're in school. Another reasonable expectation to always remember is that not every internship or co-op seems "cool," at least on paper. I had friends headed to big aerospace companies as interns, and I went to something that sounded way less interesting. While they didn't get to do any actual engineering work during their time there, I got to do meaningful engineering work, as well as grow in ways I couldn't imagine before, and that really shaped me as an early engineer. As we will continue to say, when you're getting into your career, any experience is good experience. You should absolutely chase after things that interest you and start working toward them, but don't limit yourself to only positions that seem cool on paper. As it pertains to your job hunting, especially in school, if it relates to your major, it's in general worth pursuing. You also won't know what you like until you get the opportunity to work with it some, so we encourage you to let your experiences guide you.

Now that some expectations are clear, let's talk about the act of landing a job. For the purposes of this portion of the book, we'll treat it as just two components: The resume and the actual application.

Resumes are a bit of a weird artifact in that everyone knows what they are and that they need one, but far fewer people know what makes a good one, or how to craft it. There are volumes written on this subject, and we don't claim to be experts on everything resume-related. In general, your resume needs to be a clear and concise advertisement for yourself, highlighting your education and achievements. Your school likely has some resources for crafting a good resume, and even if they don't, there are more than a few online resources to pick from. We give some of our favorite resume tips in back of the book, so you can even start there. No matter what, your resume needs to tell your story, and it needs to tell it as well as you can. That means words need to be chosen specifically, and you need to have as much supporting evidence as you can provide it. There are a lot of subjective things in a resume, and only a handful of absolutes. One of those absolutes is you 100% shouldn't lie on your resume, so don't think that is what we mean here. Rather, there are words which carry more weight and power when used, and we call those "strong" resume words, and they can be found with online searches, as well as other resources you may have available, like from different departments at your school. A resume must demonstrate that you, regardless of your education or experience level, happened to what you are involved with, instead of it happening to you. Did you make an impact? Tell us. Did you improve something? Tell us. Resumes are your time to describe your accomplishments. Just like nobody is ever going to be as concerned about your career as you are, nobody knows how to tell your story better than you. Good resume writing is a skill that will help you at every point of your career, so you may as well start practicing early and often, even if you're not quite at the point to need one just yet.

The next component necessary for landing the job is the application, and that actually begins a step farther back, with finding a job to apply to – you can't apply to a job that isn't there. Long gone are the days where you just walk around with physical copies of resumes and go into engineering companies' front doors handing them out. (Actually, we can't say that we've never heard of anyone doing that, but it is extremely rare). Instead, you're probably going to have to apply online on the companies' career page. This process, by the way, can take a few forms depending on what type of application it is. You could potentially have really hit it off with someone in person at say a career fair or similar, and they could invite you to apply with a special code or referral that skips some of the preliminary work of the application, or you could have been referred to a position by a current employee of the company, wherein the process may be a little different as well, so we're describing the general flow of applications.

This is another place where you can take advantage of your academic institution's relationships with employers. My undergraduate institution hosted career fairs in both spring and fall semesters, and they welcomed hundreds of employers each time – for most of whom the school was already a regular source of talent. This has, like everything, benefits and limitations. One benefit is that these are generally employers geographically close enough to your school that if you take a position, you likely aren't moving across the country necessarily, though it does happen. I attended a university in the Midwest, but we had some companies from California and Idaho who recruited from very specific programs within the university. Another benefit of local career fairs is that repeat employers generally know what they are getting from these programs, so you may have less difficulty in proving your knowledge at first. One of the limitations in this type of event is that if your dream company is multiple states or even countries away, you're likely not going to have a lot of opportunity to connect with them directly. In any case, you are likely going to need to apply online, even if you luck out and land a job directly, as the online application is a way for companies to track necessary data, like your contact information and eligibility to work in that country.

As you start looking for jobs to apply to, there are three major resources that come to mind for us, but you may have some more locations you're thinking of that we don't consider. First, you have the company website itself. For example, if you are really interested in working for Company X, you would simply go to the career website for Company X, and search for internships and co-ops (pro tip: If you aren't sure where to look, these are almost always co-listed as "entry level" positions, along with their positions for recent college grads). The next avenue for finding jobs is job aggregate services like Monster.com, Indeed.com, etc. I personally have never used these, so I can't speak for their utility or ease of use. The third, and what I have used most personally is social media sites, like LinkedIn.com as an example. This allows you to create a profile that really serves as a dynamic, living resume, and greatly simplifies your networking. The real beauty here is it allows you to search for jobs on criteria like location, title, experience level, and others. So, you could for example search "Aerospace Engineer" with the experience level set to entry level, in the location of Anytown, USA, and see what shows up within some radius of that location. One of the biggest advantages I see here is that you can quickly find the types of jobs in and around certain locations. For example, you may find out that Company X is in one city, but a direct competitor making a similar product, Company Y, exists just 50 miles away, so you can look at locations within industries and build a plan to gain experience. You'll also likely start building out a mind map[1] of what is where, noting that a lot of industries tend to have "hubs" where a lot of similar companies share a region, like California's Silicon Valley being a historic hub of the tech industry within the US. Hubs are common, but they do change, grow, and evolve over time. Now many of those same companies can be found in additional hubs all over the country.

A technique I (John) use often is using social media sites like LinkedIn to identify the companies in a given area, and then check the company website to look for jobs, because the jobs shown on third party sites are always subject to change and delay, so while they are often pretty accurate, they have some possibility of not showing the most current data. Needless to say, finding a job can be a marathon, not a sprint, so it's extremely helpful to know where to look. And again, we're only outlining the most common ways people find jobs, and we do acknowledge that sometimes the stars will align, and you luckily being the right person at the right place and at the right time can happen.

Once you know what you are looking for and where to look, you will start looking at jobs matched to your skill level and education. Because for a co-op or internship you're going to still be in school, a lot of these jobs have similar requirements, depending on industry. For example, if you're a civil engineering student, your internship likely won't be doing C# reviews at a banking software company, but a lot of civil engineering internships will look for similar skills and education, whether they are making bridges and roads, or doing commercial building design. This makes it easier for you as you're looking as a student, because while you probably have an idea of what you're most interested in, you can't know what you really enjoy having never done anything in the field. It's ok too if what you think you like changes as you go through school and gain experience. That's called learning and is why you went to school in the first place. It's also ok to come from the other side and have things that just don't interest you, no matter what. You know what gets you excited, and what was the reason you chose to pursue engineering originally.

If you've never gotten a job like these before, you may be surprised to learn that the recruiting for positions can take months; summer internships tend to get posted anywhere from January to typical spring break time, depending on the company, so make sure to leave yourself plenty of time and plan to apply early. Co-ops tend to start recruiting somewhere around the midway point of the previous semester, so if you are looking to have a spring and summer co-op, start looking for those postings to show up somewhere around October first.

These dates are really approximate and are meant to give you an idea of when to start looking. I knew people who landed buzzer-beater shots at getting co-ops and got them a couple weeks prior to the end of the semester, and some who were hired nearly a year early. I (John), for example, connected with a guy who would end up being my manager while I was an intern my senior year *on the drive* to start my co-op in my junior year. We kept in touch, from that January all the way to the following March when he offered me an internship – that's 14 months after I connected with him, and that connection really began three months before that, when I met his colleagues at an on-campus career event, and they started the process of connecting us. Things can and often do take time, and that's ok. You're not on anyone's clock but your own. **Your job until you land your first job is to a) qualify yourself to get it, and b) enjoy the ride as much as you can**. This will remain true when you start looking at work after graduation too. We'll be addressing both internships/Co-ops and full-time work in the following chapter.

Jobs aside, college can be a time that really sets up your future career, not just academically. If you play your cards correctly, you'll come out of school with not only the knowledge you need to be successful, but also some of the experience and relationships that can help you get there. We know that it can be an extremely stressful time as you work through various classes, subjects, and projects, but we want to make sure you keep things in perspective. Difficult times come and go, and like other difficult times before them, they will also pass. If you find yourself struggling to succeed in a certain topic, or just need some help getting through a rut, we address some of those common concerns later in the book.

If you're a nontraditional student, your college experience may look pretty different than what we outline here, and as we've said, that's totally okay. There are a lot of things we outline in the following chapters that can help you maximize your time in college, no matter what the experience looks like for you.

CHAPTER SUMMARY

Looking back, engineering school is largely what you make of it, and with a well-thought-out plan you can make a whole lot of it. Spend your time wisely between your studies and your extracurriculars, but don't let yourself get too caught up in the hustle. With planning, it is possible to enjoy the ride, *and* build yourself up in the process. All the work you are doing is *hard* work and you absolutely should take the time to breathe and recognize where you are now compared to where you started. We have provided a basic map of where you can go and what you can do within the few short years you spend in school, but it is up to you to chart your course. Below are a few key takeaways from this chapter.

- You get what you put in. Make a concentrated effort to utilize the tools and opportunities available to you (clubs, career fairs, guest lectures, and networking)
- Your GPA matters most in the first couple years after graduation; do not let a lack of motivation kill the momentum. Falling behind can happen fast and recovery can be slow
- Don't be afraid to broaden your horizons and participate in opportunities not traditionally associated with your major (e.g., being a Comp Sci major on the Formula SAE Team)
- Networking will never be easier than it is during your college years. Expand your circle to include peers, professors, and industry professionals. These connections can turn into career opportunities down the road
- Internships and Co-ops are not required to land a job after graduating, but they will give you a massive foot forward
- There is likely no better time in your career to get your hands dirty and start learning your trade, so don't be afraid to take some chances and try out things that interest you

Remember, any resource left unused is a resource wasted. The university will have tools to assist you in maximizing your position across all these bullet points, so seek them out and do not be afraid to ask questions of those who have been there and done that. Most of all, remember that in the grand scheme of things your collegiate years are short, so do your best to enjoy them without wasting them.

NOTE

1 A mind map is a technique of organization where you create a mental hierarchy showing how the components relate to the whole.

4 Landing Your First "Real" Job

You did it: The end of school is just over the horizon, and your hard work is starting to feel like it was worth doing. As we've said before, this is a great time to make sure you celebrate the huge win of graduating, and also to reflect on what you learned. For example, did you finish the engineering program you started with, or did you change majors and shift some as you learned what you liked and disliked about each engineering discipline? That's okay, by the way. You've never done this before, so learning is something that will happen every single day. Do you have a better idea about the kind of job you're looking for, and the industry you want to be in? Ideally yes, but possibly no, and even that's okay – we'll get to that too. You're about to realize the whole reason you went to school in the first place, and that alone should be quite exciting.

Of course, from here the logical next step is looking to find a job, and in my (John's) experience, there are two major camps for people looking for jobs at the end of school: Those who secured jobs way before the end of school and those who didn't. I knew both kinds during my time in my undergrad, and I was definitely in the latter category (either unfortunately, or perhaps fortunately for you, since I got to experience that for myself). I was, in fact, among the last of my peers to land a job and one of the only ones who had to wait until the next year since I was a December graduate. As it so happens, February can seem like a really long time after the middle of December.

Before we get to the true end of your undergraduate experience, let's back up some. Your first engineering job was possibly long before this point, and your first *job* was likely farther back still. If you landed an engineering internship or Co-Op (from here forward, "internship" will refer to both internships and Co-Ops for brevity, unless it is a point that needs this distinction) as part of your education journey, you've already done some of this work, and if you haven't, it's a great time to start. Wherever you are in the process, it's important to know a few things:

1. It will likely take some time.
   a. You're going to have to put in the work, and sometimes it won't seem like a linear process
   b. Don't get discouraged, as things can change quickly, and you're never more than a day away from a potential call or email that can completely change your perspective on job hunting
2. As much as you would like it to be different, the first job is often a numbers game
   a. This is true either as an intern or a graduate looking for full-time employment
3. You are going to need a plan for how to make this happen

DOI: 10.1201/9781003510901-4

As we mentioned in the previous chapter, you're going to want to get a job at some point, or else you wouldn't be in engineering school from the beginning. Ideally, you have some experience under your belt before you graduate, but if you don't, that's fine too. The truth is, every one of us had a first job, and the process is very similar with both internships and full-time work, which is why we outlined the process in the previous chapter. Let's suppose though, that you didn't land an internship and now graduation is drawing near, or possibly even behind you at this point, and you're starting to worry about the future. First of all, this is extremely normal to feel, so don't panic and don't be too hard on yourself. Even if you were able to land some experience prior to graduation, you'll still need to translate that into something you can use in gaining full-time employment.

Like with everything we've done so far, let's break this down. There are essentially two groups of two possibilities each, and everyone leaving college lands into these: (A) You have either had an opportunity to gain some experience while in college, or (B) you haven't, and from there, you have either (C) landed a job by graduation or (D) you haven't. Type "A" does not automatically guarantee you'll be a type "C", just as "B" does not automatically guarantee "D" as an outcome. We talked through some steps you can take to put you in the "A" category in the previous chapter, so here we'll focus on "B": Those who didn't get experience in college.

There can be all kinds of opportunities to improve how you look on paper while in college: Improving your GPA, taking more or different classes, becoming involved in student design teams etc., but the closer you get to graduation, the less you can do to change it, with the remaining opportunities ultimately going to zero. Put differently, once you're done with your undergraduate experience, you have what you have. That may sound like an imposing negative thought, but it's just a fact. To that end, I (John) actually knew a guy during my undergrad who was on scholarship and double majored in engineering and something else for the sole reason to help keep his GPA up – if you're creative, willing, and have some opportunities, there are a lot of things you can do to help shore up your GPA.

What it means in practice is you are going to have to really focus on how you present yourself with your resume, networking, and job-seeking skills. GPA often becomes the main driving force in landing your first job, so as a general rule of thumb, the lower your GPA, the less picky you are likely going to be able to be in your search. A lot of the biggest names in respective industries want high GPAs for their early career and recently graduated folks, and a lot of them even require GPAs as high as 3.7 or more out of 4.0 to even be considered for a position. As someone (John) who had a sub-3.0 – I graduated with a 2.9 overall because I really struggled for a while for reasons we'll get into later. Fortunately, I had turned the ship around during my last two years and my in-major GPA was still much more favorable – I knew that wasn't a likely occurrence. If you also have a low GPA, there is some good news for you: Just a few years out in the professional world, maybe within three to five years after graduation, it is very likely that nobody will even ask about your GPA again. If you decide to pursue a graduate degree, then you also have a chance to improve it even further, and I can tell you from personal experience that when you have a Master's degree with a better GPA than you got in your undergrad, your undergrad GPA becomes nearly meaningless and nobody will even care. Still though, a graduate degree is not the right answer for everyone, and it not being an issue in five years doesn't help you at all when you are graduating and need to land a job soon, so what can you do?

If you haven't been doing it to this point, now is *THE* time to pull out all the stops and use every resource you have at your disposal. If you are still on campus, take advantage of everything your school offers related to job hunting: Resume help, career fairs, networking events, or literally anything else they have. If they don't have enough or if they don't offer anything at all (which is rare), then you're going to need to find ways to make it happen yourself. This can be things like your instructors and their connections, professional societies, social societies, etc. and it's time to start working to get yourself ready to apply to the jobs you can find for which you're qualified.

Let's also go back now and address one thing we mentioned just briefly at the beginning of this chapter: Do you know where you want to land industry-wise? If you have had an internship, or possibly done some student research/on-campus work, then you may have an idea of what you really like – for example, you may know that as a Computer Engineering major working as a Software Engineer Co-Op you got extensive exposure to MATLAB in a testing environment for the heavy trucking industry, and you loved it – or that you absolutely hated it and never wanted to touch it again. In either case, both are great pieces of information because with experience you learn what you like and don't like, so this job experience is a big advantage for more than just the obvious reasons. If you didn't land that job, you don't necessarily know what you're in for with certain things, and that's okay too. By the time you're a senior nearing graduation, you likely know at least some of what you like, say in the same example you're a computer engineering major and what you know is that you really like Python-based development, and you absolutely loathe doing anything related to Fortran. This is still good information to have as you know you probably won't like jobs whose main function is Fortran scripts. The point here is that information is important to you, and a key takeaway for *any job* at *any time* in your career is that there are levels of quality of information, but you likely will *never* have perfect information available to you. With that, you're going to need to make an informed decision and pursue your best judgment when applying to jobs but do so with the knowledge that jobs aren't permanent. The best jobs and the worst jobs have that in common – they aren't permanent. You can take a job that meets most of your requirements and learn from there, and you can always find and take another one later once you have more experience.

Working with the experience you do have, whether that is from working during college, or just the experience you gained from your classwork, you know a lot about what you do and don't like as it pertains to the type of work you're looking to find. Different examples of this could be you're an aerospace engineering major and you know you're way more interested in the propulsion side life than you are in the aerodynamic side, or atmospheric flight versus space flight, or perhaps you are a civil engineer who knows you want to be working on buildings instead of infrastructure projects like roads and dams. All of these are great pieces of information to have, and the reality is a lot of this will come organically as you work through your curricula on your way to graduation. The other case of unknown unknowns[1] here is if you really like something (the specific work) but you aren't sure where you'd like to do it (the industry or possibly the location). Say you were to absolutely enjoy mechanical design, but you aren't sure if you'd like to put that toward consumer goods, or parts for a race car. In our opinion, this is actually the better of the two places to be, because it opens up so many things about your job search. If you are dead set on working for a team designing race cars that's great, but there are only so many race car teams in the world, so some probabilities come into play. If you just want to hone your skills in mechanical design, there are far more opportunities out there across multiple industries. We always encourage you to follow your dreams and relentlessly pursue what you need to make them a reality, but we also want to be realistic with you up front and say that landing a dream job out of college is a rare event, and not necessarily the norm you should expect. Even so, it does happen.

So what? Where does this leave you? Ok, you're about to graduate college, and you have a notional idea of what you want to do either by work experience or just finishing your degree or, most likely, a combination of both. You know the GPA you have is important but not the only thing in the world, and you know that landing a "unicorn" job right out of the gate is not necessarily going to happen to everyone you know, you included.

Most people now will tell you that this is the time to "network," but if you're like we were around graduation, you don't really know what this means. Even if you have an idea of what it means, maybe you aren't sure how you can network with people in the industry, since you feel like you don't have much to contribute. Hopefully, you were able to learn and utilize some great skills during your collegiate experience (do design teams ring a bell?) but maybe you weren't able to, or you came into this a little late. That's okay too. Life happens and we get that, so don't feel like you've already missed the boat. Regardless of when you start, you're going to be starting from scratch either way. Networking can seem a little nuanced and out of your comfort zone at first, so we can guide you through that. Let's say you're still in school and as a benefit of being in one of those design teams/school clubs we mentioned, you get to meet a cool engineer in a field you're interested in, Ms. Jones. You meet Ms. Jones after her presentation to the team, and she says she really enjoys helping students find their passion and land jobs, and that her company loves to bring on graduates of this program at your school. From there, you go to one of the professional social media sites, you find her profile and you send her a connection request along the lines of

> Hi Ms. Jones, My name is John Rogers and we met at your presentation to the design team at University X last week. I was really interested in the topic you presented, and I'd love to keep in touch for future opportunities and positions.

You connect, and she directs you to apply for a job with her as your internal referral, and just like that, the company hires you when you graduate. Simple, right? If there were such a thing as an ideal way in an ideal world, that would be it, but as we said, life isn't always ideal.

Let's now propose the probably more likely scenario, where you have never met Ms. Jones before, you don't know what her stance is on students, and you have no clue what kinds of people her company looks for. What you do know is that she has a job in a company that you're interested in from doing your research, and you see that her educational background is a close match to what you are doing. You send her a message out of the blue, but in this case, you say something more along the lines of

> Hi Ms. Jones, My name is John Rogers, and I am a student in Systems Engineering at University X. I see you work for Super Cool Company, and I'd love to ask you some questions about getting a job in that industry.

Now, you don't know how she'll respond, so you're hoping she does in good faith, and it is my experience that most (certainly not all) people in engineering positions are willing to help the next generation of engineers at least some, so they will generally respond, but the level of helpfulness and responsiveness you get from these interactions will vary.

This provides us an excellent segue into some networking best practices, or outside of that lingo, some "dos and don'ts." First, and we cannot stress this enough, DO NOT just email someone you don't know and say any form whatsoever of "Please hire me"/ "Find me a job" or anything that sounds even remotely close to this. This person doesn't know you, and it's horribly very rude to ask someone to go find things for you or to refer you to something within their company without some groundwork. In fact, giving someone an errand to run is bad practice even if you do know them. Universally, please do not just ask people to go find you jobs. Second, in the event Ms. Jones responds, it's good to limit your initial inquiry to a few questions, ideally no more than two or three. You could add these to your initial message, then you save a step of back and forth if and when she does respond. For example, after you say you're interested in getting a job in a specific industry, you could ask "what was the most important programming language you needed to get there?", or say "I am pretty proficient in Dassault SolidWorks, but I heard in your industry, they predominately use Siemens NX, and I was wondering if you were able to learn that while you were still a student?" Both of these are examples of perfectly relevant questions to ask, and it's even ok to ask a few questions at a time (again, no more than probably three), but you don't want to make it an arduous task for them to respond, or worse, make them feel like you gave them a chore. The third tip we'll give here is that you should match your inquiry to those relevant to your search. If you just reach out cold to the CEO of a Fortune 500 company, more than likely they won't respond to you. The director of engineering likely won't either and is probably not the best usage of your efforts.

There is always a chance that those closer to your own experience level won't respond in the first place either, and that's okay. You shouldn't expect a 100% response rate, but asking good questions to relevant people will greatly help your chances. (Pro Tip: use filters on social media sites like LinkedIn to find people with similar backgrounds as you, like "I see you also grew up in Florida," or "from one Auburn Tiger to another, War Eagle!"). I (John) have had some really good conversations just by reaching out to people I didn't know and asking questions using the above "template," because most people remember they started somewhere too, and they don't mind taking a few well-intended questions, especially if you have done your homework and you share common ground. We encourage you to always be open to conversations and helping your fellow engineers, because one day, you'll be the one the new engineers come to for help, no matter how far away that may seem at the present moment.

You'll see there are all kinds of networking opportunities available to you, once you start looking for them, whether you are in school or not. Professional societies often have regional chapters and local speakers at dinners, meet-and-greets, and other events, and they often have discounted or even free events for college students and recent graduates. Other societies to consider are alumni associations of schools and departments, and more general professional societies like Toastmasters or Young Professionals.

There are even opportunities you wouldn't normally come across, once you start thinking outside the box, like asking your mechanic or hair stylist if they have any clients who are engineers or students. Networking is a world of possibilities, and the ways to open these lanes of communication are virtually limitless, even if you are off campus or into your career. Hopefully, none of them are intimidating to you, even if you are more on the introverted side of the scale. I (John) am an extrovert and that helps me immensely, but these methods do work for the introverted among us. If you find yourself intimidated by the thought of asking complete strangers for some information, rely on your common ground and focus on that. Not ready to ask a full-fledged three-question inquiry? Start with "I see you also are an alum of the FSAE program at University X" and go from there. All these networking relationships can be built over time, and in fact should be built over time. You're not trying to make a best friend in a day, but rather gain some insight into the company/industry/job you are looking at, and gradually gain some exposure to those who are already there.

Ultimately, the goal of networking is for you to get yourself out there where you know someone in industry. You don't even necessarily need to know a lot of "someones" in a lot of industries, but rather the right people in the right industries for you. Quality is absolutely valued over quantity, but likely you'll need to grow one with the other. Why is this the goal in the first place? Because as you gain exposure to industries and the people in them, you will find more companies, industries, positions, and opportunities that fit what you are looking for, as well as hopefully some people who can help you get there. This is not to say relationships are transactional, but rather that we all need people in our circles at various points in our lives; you are going to need some people in your circle, and one day, it's very likely that someone will need you in theirs, and it helps to be in a place where you can do something about that.

Networking is intended to give you not only a good network of people you can help and be helped by but also to help your chances of landing jobs, because it can be a very competitive space out there. We are certainly not telling you that if you network "right" you'll never have to get jobs the traditional way, or that knowing the "right" people is going to get you through life, but rather that there is definite benefit to not having to go at everything alone. In the job I (John) had when we started out with the writing of this book, I actually got it because I interviewed for another position within the company, and while that hiring manager couldn't take me on, he referred me to another hiring manager within the company, and about two months later, she was the one who hired me. I certainly still had to interview and follow the standard process, but how much better is it if you can start off knowing you will have an interview that came as a direct result from someone else's vote of confidence in you? I didn't even know I was doing it at the time, but I was growing my network as a direct result of the knowledge and performance I had to that point.

Seeking and landing jobs is a critical part of your career, and you're likely going to do it more than once, so it's really important for you to learn and get comfortable with the process, as well as taking any advantages you can along the way. Networking is a piece of the puzzle, but not the entire puzzle, as you've likely seen by now. To land a job as we've laid out to this point, you will need skills, you will need a position to apply to, and you will need a way to show these skills. One way being networking, as outlined above, and the other being a resume, which we briefly introduced in a previous chapter and return to now.

Through this whole process, questions will likely arise, and these are among what we'd expect are going to be your most encountered ones:

- Where do I Find Job Postings?
- Should I Shoot for the Job of my Dreams?
- Is my Resume Good Enough?
- I Landed an Interview, So What Now?

We briefly addressed where to find job postings in a previous chapter, so we'll finish answering this question now:

If you run a quick internet search query on "engineering jobs in my city" without a doubt you will be inundated with what seems like way too many options and no real clarity on which sites are legitimate. What you will find is there are three major approaches to finding a job posting: Job posting aggregates, direct postings on company websites, and professional social media/networking. Each of these options has their pros and cons that we will delve into.

The first and most diverse option is job aggregates. Aggregates such as Indeed, CareerBuilder, Glassdoor, and LinkedIn Jobs are a seemingly infinite source of potential jobs. Websites such as these allow you to filter and scale your search with key parameters such as:

- Job Type – Full-time positions, Contract, Part-time, Internship
- Location – City/State in which position is located
- Experience Level – Entry Level (0–3 years), Mid-Level (3–6 years), Senior Position (7–10+ years)
- On-site/Remote/Hybrid – How often will you be in office for the role, if at all
- Salary Range – Usually by minimum salary per job ($60,000+ for example. Pro tip, if you aren't sure what the salary range is, several states like Colorado and New York now require the job salary to be posted, so you can search for similar roles in those states to get an idea. Of course, you'll need to adjust the cost of living to the areas you are looking at to have a more even comparison, but it's a much better start than speculating)
- Education Level – Bachelor's or Master's degree (or beyond)
- Posting Date – The newer the posting, the better visibility your application will have – to a point

Utilization of these parameters will result in a more informed job search which in turn results in an easier path to getting hired. It is not in the best interest of the hiring manager nor you as the applicant to waste time on positions that don't fall within your desires and qualifications. You may be wondering which of these aggregates is best or if you should make accounts for every single one to maximize visibility on new openings. Each of these has its upsides and downsides but two in particular stand out to us: Indeed and LinkedIn.

When you create an account through Indeed you are able to fill in their preset questionnaire that is forwarded directly to hiring managers of positions denoted as "Easy Apply" and if the boilerplate questionnaire doesn't cover more specific requirements the questions are prompted upon pressing apply (typically just asking how many years of experience one has in specific programs or fields relating to the position.) Additionally, Indeed has salary and benefits expectations for a large quantity of positions, which helps filter out any time-wasters. And let's be honest, very likely the further into the job hunt you are, the less patience you're going to have for delays. Indeed will also keep track of which jobs you have applied in the order they were applied to. For anyone who has submitted dozens of applications a week, it can be stressful trying to keep track of which one is which when a hiring manager or recruiter calls to get more information or set up an interview. With this feature it is possible to find the company and really give the job description a thorough once over to better prepare yourself for the conversation ahead, a subject that will be touched on further along in this chapter. Last, recruiters are able to review your Indeed resume and submit openings directly to your inbox that they deem you qualified for.

LinkedIn Jobs is similar to Indeed in terms of search parameters and volume of job postings, but there are certain key differences that justify utilizing both of these resources when job hunting. LinkedIn sets itself apart by being directly connected to your professional image you put out for the world. This adds some personality to your applications. Recruiters can see your qualifications on your resume and immediately click a link to see your professional headshot as well as any certifications or professional accomplishments you choose to highlight on your page. The recruiters can also directly message you through LinkedIn's chat feature for rapid communication. It should be noted that though it is very similar to your standard social media messenger (in fact, LinkedIn is often classified as a social media website), be sure to maintain a professional demeanor through all communications. Formalities should not be forgotten, though you can match the tone of the recruiter after a couple messages. LinkedIn also allows you to save a resume for quick attachment to jobs that are posted with "Easy Apply." In addition to the features listed above, one of our personal favorites is the ability to click on the company's profile through the job posting. This allows you to dive into their social media and see what new and exciting things are happening as well as gain insight into who is in charge of various departments (potential bosses) and overall company size.

A pitfall with either of the above methods is they aren't absolutely guaranteed to be accurate postings all the time. They will be close, but you'll occasionally run into the case where you see a job to apply to but when you try to apply, the posting has already been removed by the employer, so just be aware of that as you search, and always treat the company's own site as the accurate one.

The second option for finding job postings is directly through company websites. Have you been dreaming of a career at your favorite automaker since you were 11? Well now that you have obtained that elusive engineering degree, you type their web address in and after a little bit of detective work you find the careers page with every opening in the company listed and available for you to search through. For large companies this is a fairly useful option. Their search engines are not too different from those of Indeed or LinkedIn, with the exception of being relatively narrow in selection.

As engineers who have applied to engineering jobs in just about every format available, we can say direct postings are wonderful if you are only going this route for a handful of companies at most. We say companies and not positions because of the fact that most major employers (Fortune 500 level companies) run on similar software as each other for their hiring process. With that, after filling out your applications it is rather easy to select any multitude of other positions that you are interested in and have all of the information you so carefully entered appear instantly and pass your resume on as you see fit. This is great but can have its own major pitfalls too.

One problem with applying directly to major companies is their application processes are usually very tedious. You will be asked to upload a resume and cross your fingers that the format is readable by their reviewer because if not, you will likely be writing everything that is in your resume, word for word, into the application software. For that reason, we recommend you upload your resume in either a Microsoft Word or PDF format only, and the application should declare which of the two they prefer, if they prefer one over the other. In our experience, this is *usually* PDF format, which would be our recommendation. This detailed application process can take upwards of 30–45 minutes for some of the more rigorous questionnaires, and if you're applying to certain positions within the US government, it can take upwards of two hours or more. Now take that time and multiply it by the number of direct postings you are applying for and it is easy to see how a day of applying for jobs can seem to have evaporated and gone by the wayside. On top of this, there is usually not an easy form of communication between you and the company like there is with the other websites. By the end of it, you will end up with another username and password to remember for an account you will only access for applications to that specific company, yet you will still be waiting for a response via phone number or email that you provided in the application. Again, if it needs repeating, these things can move pretty slowly so don't be discouraged if things take weeks. Although there are certainly times when companies call you back the next day, we've found getting a response in even a week to be quite a rare occurrence.

By all means, apply directly with the company, but for a diverse job search, more success is found through Indeed and LinkedIn, which brings us to the next question you'll probably have: Should I shoot for the job of my dreams? You may find this question popping into your head more frequently the longer you are on the hunt. Perhaps the hiring process of your dream firm is long and arduous, or at the career fair your senior year you find out there are only five positions available but sixty engineers ahead of you just waiting to talk to who they hope will be their future employers. It is easy to get discouraged along different points of the job search, and we know that to be the case ourselves.

There is no exact science for answering this question as it is entirely based on the reality of an individual's situation. The biggest question here is simply do you have the means to wait it out? Fresh graduates may find their student loan payments coming due or they need to move out of their student housing. Maybe their old car from high school is finally on its last legs, or possibly all of those situations and more. The same recent graduate may have barely made it to graduation with a less-than-stellar GPA and no extracurriculars. We hope if you are reading this book while you're still in school that you take the correct steps to set yourself up for success early while there is still time to do so, even if it's uncomfortable. You'll certainly thank yourself later if you can right the ship now. Still, others may be in the early years of their career looking for a change with no hard timeline to transition. Each and every one of you reading should take an honest look at your situation and find that answer for yourself.

Having said that, we are still going to give our opinion on the matter. Early in your career is an interesting place to be as whatever position you end up taking will help guide your future in a multitude of ways. You will learn the ins and outs of what it is to be an engineer on a day-to-day basis and be exposed to many aspects of the career. For example, Mechanical Engineers have the option to be manufacturing engineers, or test engineers, or quality engineers, or product design engineers, or virtually anything else amongst other subsets. Each of these categories of engineers will find themselves working alongside the others in some small and most medium to large size companies. Though hired as a test engineer, one may find themselves drawn to design or vice versa. Unless exposed via internships or co-ops, it is hard to get a feel for what each of these positions might mean. Additionally, you may be hired into an industry and have one opinion that completely shifts after a year or two of exposure to its inner workings. You don't know what you don't know, and that's completely okay. More than that, it's normal to not know exactly where you want to land.

While you are still in your early career it is easy to pivot, and your dream company might just want to hire at the two-to-five years' experience level much more than they want to hire fresh engineers. There will always be an opportunity to find your career home and it is our shared opinion that you should explore many options when it comes to your career. Interests change, goals change, we as people and professionals will change. So, take some time to think about what is driving you to be an engineer for the NASAs or Intels or Boeings or Procter and Gambles of the world. Is it the joy of being a part of a company you love? Is it the potential for larger paychecks offered by the larger firms? Or is it the prestige of the title?

There is not necessarily a wrong answer as long as it is your choice because it is ultimately your career. It's okay to want more money because money is a tool for you to use; money is a means to an end. We've known engineers who made more money and hated the job they had, and we've known engineers who made less money and loved the job they had because of other factors. Your career is your career, so as long as you're getting what you want from it, you're making your career work for you. After identifying your driving force, you can take an educated stance on the job market and what is available to you. You may find that there are positions that are adjacent to your dream job. Particularly in the space and defense sector, you will find the big-name organizations will often subcontract parts or segments of their work to other companies you probably never knew existed. There are options beyond what you may be pigeonholing yourself to, so remain open to looking around. Now, are we discounting your dream job? Not at all, and you should emphasize on the positions you find yourself most drawn to during your search. However, you must keep some perspective on "settling" for what you are able to lock down on a reasonable timeline.

Next, we know you can't find a job without a resume, which is again, kind of a necessary evil. Or rather, they seem like a necessary evil when you aren't sure what happens with them or where they go, as long gone are the days of applying in person. Fret not, we'll get to that next. A lot of fuss is made about ATS' role in your job application journey. An Applicant Tracking System (ATS), is used by most companies as a database for the resumes and applicant information they receive because as you can imagine, a Fortune 500 company gets a *lot* of applications each year. The fuss comes from applicants believing the ATS is the first set of "eyes" that reviews their application and is what kicks it out, sometimes nearly immediately, while numerous people on the recruiting side say it does no such thing and is merely a database. We aren't on the back end of the many ATS systems out there, and we don't claim to know what companies do or do-not-do with their systems, so we won't speculate on the specifics of what is done.

What we will say is first and foremost, your resume absolutely has to do a few things; primarily among them, be readable by a human. There are all kinds of gimmicks and tricks out there you can find with a quick web search, and truth be told, we've tried some of them ourselves in college to no avail. My advice to you is to make sure your resume tells the story you want it to tell, and to make sure you highlight the relevant experience that a position needs you to have. If you are applying to a job that requires 10 years of Python experience, and you just started learning C#, your application probably won't get too far, just as if a position requires an active DOD security clearance to start but you don't have one, your chances are going to be pretty slim.

We said resumes can seem like a necessary evil if you don't have a good handle on them, so how should you view them and what do you actually need? First, a resume is a tool for you to highlight and present your platform, and why you are a great candidate. Once you decide to start viewing a resume as a tool, it becomes far less intimidating, and you can view it in the light it deserves. As for what you must have in a resume, you're going to need to highlight the education and experience you have, and how that makes you a relevant fit for a position. (Pro tip: If you are just starting out and have very limited to no experience, you're more likely going to list your education first, whereas if you are out of school more than two years or so, you'll list your experience first in the resume, and your education later. We included a sample resume in the FAQ portion of the book, Things we wish we knew sooner, so you can get a feel for what we mean.) Also, if resumes and resume building intimidate you, do not feel like you're alone. We hear a lot from early career engineers and other professionals that nobody ever taught them to write a resume, and looking back, I (John) think nearly everything I learned about resumes was self-guided, so don't feel bad if you didn't come out of engineering school feeling like a professional resume writer. We could not name anyone who did either, and we've had to learn as we go.

Your resume needs to be concise, yet descriptive. Again, we are never telling you to lie or inflate your resume, but this is not the time to be timid. Did you do something that was brilliant, save someone a bunch of money, or mentor a lot of people? Tell us. The most important thing to remember is your resume has to be legible by a person, no matter what you may have heard about ATS's role. This means your resume needs to follow some general guidelines: if you're starting out, it's best to keep it to a single page (two pages is ok for later on when you have some experience), stick to normal "boring" typefaces (fonts) – like Arial, Times New Roman, or Garamond – use between 10- and 12-point font size, and treat page space as precious, meaning if it doesn't help prove why you're a great candidate, it should not be there. Make sure you have no spelling or punctuation errors, and keep the tone to a professional, active voice – "drove project to completion" is a lot more active than "helped finish a task." Generally speaking, if you are a student, only list your GPA on your resume if it will help you stand out, like a 4.0 or close, and if you're a graduated professional, leave it off the page entirely. If the company wants you to list it, they will ask during the application process, and, eventually, it won't even matter. Once you have some good experience gained in your career, GPA loses a lot of its value to employers.

As you work through the process of applying, you'll wonder if your resume is good enough. Once you've utilized the preceding information we've given, you become faced with a job posting you are qualified for, in your area, and that piques your interest. So, what next? How do you know if the job you are interested in will be interested in you?

One of the biggest mistakes one can make during a job hunt is creating a broad resume covering your entire experience and submitting it blindly to every job posting you find. Tailoring the same resume to align to a litany of job postings is a bit of an art but when done right, it will catapult you to the top of the candidate list in the eyes of recruiters. Imagine you are looking to hire a propulsion engineer and you come across a resume loaded with information related to Heating, Ventilation, and Air Conditioning (HVAC) design. Is there relevant experience hidden within? More than likely there is, though the impact it has will be minimized by the lack of focus. Hiring managers and recruiters are concerned with these key features on a resume:

- Relevant Experience
- Tangible Results
- Skills
- Education (with relevant classes highlighted)
- Contact information
- The speed with which they can read the resume

All pretty logical, but in the effort to encapsulate all that we are capable of sometimes we showcase the wrong information. What do you need to do to properly highlight these features on a resume?

Let's start with the most important aspect of a resume, your experience. Experience can be difficult to write about for early career engineers and fresh college graduates. It is easy to sit there and wonder "How do I even detail my professional experience when I haven't worked as an engineer yet?" This is where resume writing becomes a bit of an art as well as a science. Your first step is reviewing the job posting that you are aiming to apply to. The format for these is typically an opening statement about the company or position available that loosely defines the company as well as the responsibilities of the position. Following the introduction are the job requirements and then the desired qualifications. The difference being job requirements are bare minimums to apply and desired qualifications are what an ideal candidate will have. Be aware that some companies have compliance policies that dictate experience level, so if a job says minimum two years' experience you may not be eligible with even 22 months of experience, no matter how much of a small detail that may seem. It's surprising how many employers still have rigid, archaic practices that don't seem to be moving into this millennium. This can be annoying, we know, but this is typically for larger companies and positions with higher titles than entry-level engineers (positions beyond Engineer I or II, for example). As for desired qualifications, it should be noted that no candidate is ever ideal, so don't get discouraged if you check only a few of the boxes here. If you have some experience, meeting around 80% of the requirements is a good target, as it gives you some room to grow into the role, but helps you confidently know you can fulfill the responsibilities of the job.

The next step for you is comparing the requirements to your existing resume. Do you have anything that matches their requirements? Great! Under your previous jobs, personal/school project(s), or internship(s), highlight this information by detailing in exact and concise terms your relevant experience and bring it to the top of the list. From there, you'll need to add and fine-tune your experience that may not have been included as yet. For example, a requirement such as "Ability to drive iterative design concepts while meeting project deadlines" really means you can quickly solve design problems as they pop up and pivot a design to accomplish the original task quickly enough to meet your timeline. If you participated in a senior capstone project, you have direct experience in this. The professor is your "boss," graduation is your deadline, and midyear design reviews are where you discovered issues and had to pivot. You would be surprised how much of your personal and collegiate experience relates to the professional world; it is all about how you frame it. As we said in the beginning of this book, problem-solving is one of the single greatest things you learn as an engineer.

After detailing this experience, you may think that is enough, but simply stating that you have accomplished something is generally insufficient for the majority of hiring managers. Take this for example, "I designed a process to reduce cost on an existing product." On the surface this is fantastic, you took a known entity and managed to increase the profit margin for your company. Yet there remains a lot of missing information: How much was the cost reduced in percentage? How was it reduced? Did you improve the time to assemble, material cost, make a design change, or was a manufacturing method updated to the latest technology? What were the yearly savings to the company that resulted from this change? Quite simply, tangible results need to be described in numerical values to drive home what value you can bring to a company. This will also typically lead to a talking point in any interviews that result from your application. **Hiring managers love numbers**. If you're not sure, your resume probably needs more numbers. Quantity is every bit as important as quality on resumes.

Skills are usually grouped on their own, separate from professional experience. These can be broad, such as soft skills like "communication" and being a "self-starter," but specific skills like computer software, GD&T (Geometric Dimensioning and Tolerancing, a method of both defining and communicating engineering tolerances), or unique manufacturing techniques should be accompanied by the number of years of experience you have with them. Education is one of the resume factors that rapidly decreases in relevancy the further into your career you get. In the beginning, your tech electives and GPA are what separates you from the pack of other new grads with similar experience levels, but as you gain experience on the job you will find that portion of your resume shrinking to the university you attended, what degree you hold, and unique technical electives that are relevant to the job for which you are applying. Contact information is very straightforward but it is absolutely necessary to have an easy way to identify who the resume belongs to and how to follow up. Never make it unnecessarily burdensome for a recruiter to reach you. (Pro tip: You don't need to identify things on your resume like your specific home address, your marital status, your nationality, your age, or a picture of yourself. These are all somewhat common in various other countries, but in the United States, they are unnecessary and could subject you to subconscious biases by employers. In fact, some of them are even illegal for an employer to ask you in an interview, and you do have rights.)

Combining all of the above will help form a resume suited for your needs, and the needs of the company you are wishing to join. As you explore the job boards you may see a similarity between positions in the same industry and you can more broadly tune your resume on an industry-to-industry basis than fine-tune to the posting, which could end up ultimately saving you time.

Alright, you put in some solid effort, you worked to combine all we've laid out in crafting a great resume, then you found and applied to jobs, and you finally landed an interview, so what happens next? You did it, all your hard work has paid off and companies are starting to respond to your applications. There are a few ways this process can go. The most common situation you will find is the recruiter or hiring manager will call you and ask a few baseline questions confirming your resume is accurate or maybe fleshing out some details that needed clarity. This is often called a screening or initial interview. From this point they will schedule you for additional interviews with either a singular hiring manager or a team of relevant employees/managers, called a panel interview. Most employers now use some form of an ATS we mentioned earlier, and one of the benefits of these systems is you can track your application on the company-specific job portal and some even have highly detailed timelines for how long you'll be in a given stage of the process. Remember though, this process will take some time, and may even take more time than you're comfortable waiting. Keep applying to jobs until you've landed one and have a firm start date.

A rule of thumb is the larger the company, the more formal – and thus longer – the process. We have had interviews scheduled for four one-hour long blocks split among a peer, a manager from another team, the hiring manager, and human resources. I have also had a first interview at a smaller firm with the president of the company as the only interviewer. This is not always the case, but will help set expectations. We've also had interviews that lasted eight hours and were completely exhausting for everyone involved, and we've even had interviews that weren't interviews per se, just a meet and greet with a job offer. In fact, on two different occasions, I've (John) met the team *after* receiving a job offer.

What should you expect from these interviews? How can you put your best foot forward? How do you impress the interviewers? How do you quell the nerves? What questions do you ask them?

Great questions!

While each interviewer and each person being interviewed can differ substantially, the purpose of the interview remains the same, and that's to prove yourself a great fit for their needs. Knowing what to expect from the process is a great start to feeling prepared. As mentioned above, there is the potential for a multitude of people from a variety of disciplines to be present at the interview. Many of whom do not have engineering degrees and won't necessarily care about the technical aspect of what you bring to the table but will care about other skills, including soft skills like how well you can participate as a part of a team. Prior to the interview, review the interview panel and take note of who will be present (if not mentioned by the recruiter through email or the initial screening call, be sure to ask for this information). Direct engineering managers will likely be interested in your technical skills and knowledge base first and foremost, peers typically look for a mix of technical knowledge and personality, while HR will make sure you meet hiring compliance and will mix well with the team among other factors. Preparation is key, so make sure you identify your strengths in these categories and be an honest and professional version of yourself in the interview. You don't have to be perfect, but you will need to be able to articulate why you're a good fit.

Easier said than done sometimes, right? The interview process can be intimidating for many. You spent at least four years learning everything you need to become an engineer, but so too did all of your peers. How do you differentiate yourself?

Engineers, on the whole, have the unfortunate stigma of being labeled socially awkward and difficult to communicate with (though it's our opinion that this isn't always a fairly deserved reputation!) If you are aware of yourself, this can be an advantage in the interview process. It's an opportunity to be straightforward, to be clear and concise, and to make eye contact (even in online interviews, which are becoming ever more common). Show that you can talk to new people across the professional ladder with confidence and certainty in what you are saying. Strong communication skills are a massive edge over the competition and will position you well with management at all levels. Throughout your career you will undoubtedly have to report to someone about successes, failures, and status updates on your programs, and a manager will be much happier if they can trust you to present this accurately and confidently. Additionally, if you have something on your resume, be absolutely confident that you can defend and extrapolate the claims.

Don't forget, at this point, you've earned the right to be in the interview, so you belong here. There are all kinds of additional resources to help you prepare for interviews, including career centers, internet searches, and even social media. One thing to be aware of is that many companies use some form of the STAR (which stands for Situation, Task, Action, and Result) method for interviewing, which you can always identify because the questions nearly universally begin with some form of "can you tell me about a time when ___?" A STAR question is going to set the stage for an event, what the action was, and how you handled it. For example, "Can you tell me about a time when a person on your team wasn't doing their assigned responsibility and it caused you to miss a deadline? How did you react and how did the project get back on track?" Knowing things like these are important because if you know the kinds of questions you'll get, you can have time to think about good, relevant answers to those questions. One question that commonly shows up is a form of "can you tell me your biggest strength and weakness?" People are often not great at talking about what they need to improve upon, so you'll need to think about that answer before you need to give it. Look up STAR interview questions with any internet search engine, and use this to get a feel for what you may be asked prior to your interviews.

Throughout my career, I (Sean) have interviewed dozens of times and to this day I will still feel some nerves when I'm walking (or calling) into an interview for a position I am very interested in. For some, this never existed from the beginning, for others it fades with experience, and others will be just as nervous the first time as they are the hundredth time. Nerves are good, as they indicate an opportunity is exciting and important to you, but they must be controlled. How do we control these seemingly uncontrollable nerves? Through familiarity, practice, and planning. Chances are you were not the most confident child in the world the first time you hopped on a bike, though if you were to hop on today it wouldn't bother you in the slightest (assuming you learned to ride, of course). We familiarize ourselves with these situations and move forward from a position of strength, which we get by preparing to the extent possible and by practicing.

How will you prepare? By studying! After at least four years of engineering school, we should all be pretty good at that. Start by studying the job posting a day or two prior to the interview and once more the day of. Connect the dots of what they want to what you have or do not have on your resume. It is a pretty safe bet that what they are asking for in the job posting is what they are going to ask you in the interview. Write each requirement down in the form of a question. For example, if the job description lists "experience in solid modeling with CREO" you can write "What is your familiarity level with CAD software like CREO?" and begin to form an answer that highlights your strengths and displays the complexity in which you can utilize your knowledge. Next, you will want to study your own resume, which sounds crazy, we know. Of course, you know what experience you have; you wrote the resume after all! But in the heat of the moment can you guarantee you will say everything you want to about your qualifications? It is best to have a deep familiarity of what is on your resume and what you wish to portray by having the information displayed. Finally, you can study the company. Go on their website or LinkedIn page and see what products they produce or services provide. A lot of companies are building their presence on social media and video platforms as well, so you can often find a lot of information about their products and services that way. Find out what makes them special or at the very least, what makes you interested in them, as well as what you will be doing if hired. And make sure to take notes! It is perfectly acceptable to have notes in front of you with thought-out answers, bullet points, and questions you wish to ask the interviewers. Good interviewers will always allow and plan some time in the end for you, which leads us to what typically ends an interview when they ask you, "what questions do you have for us?" What questions should you have for the interviewer?

While this may seem like a pretty minor part of the interview, three to five well-thought-out questions will impress an interviewer and may just set you over the edge compared to your peers. Detailed, relevant questions show you are engaged with the opportunity and aren't just going through the motions hoping to find whatever you can get. Some example questions I (Sean) like to ask are:

- What is the greatest professional growth you have experienced working for *insert company*?
- What excites you about the company?
- What unexpected benefits have you found working for the company?
- If I am hired for this position, where would you like me to be in the next 6–12 months?
- How is the work culture?

And some of John's favorite questions to ask:

- What does success look like in this position?
- What are the biggest challenges/barriers to success/obstacles you encounter?
- Why is this position open?
- What is your onboarding process like?
- Was there anything you'd like me to go back to and revisit or answer more thoroughly?
  - This is kind of a wildcard, and I wouldn't recommend it for every interview. If used correctly though, it allows you to get a second chance on parts of the interview that perhaps weren't as smooth as you would have liked them to be. In this case, you'll likely know when to use this question and when to avoid it. An alternate form of this question is "is there any reason you feel like I wouldn't be a good choice for this role?"

You will inevitably find what matters to you in a future employer and adjust the questions accordingly. An important detail about interviewing is you are not the only one under review. You are interviewing the company and manager just as much as they are interviewing you, and that's true of the entire process, from the first application to the job offer. If anything needs clarity or if you are unsure of any answers, do not hesitate to ask them to elaborate. A candidate with a well thought out resume and the ability to interview well has a surprising amount of power in the interview dynamic. Use it to your advantage and don't be afraid to investigate if this is the position for you. You can also know right away that this is not a good fit, and know you wouldn't accept an offer should one be made. I (John) have done that before, and I knew I wasn't going to be working at one place literally 5 minutes into an interview, when a very highly inappropriate joke was made. I politely said that was inappropriate, though I did finish the interview. Trust your gut because those things you know aren't a good fit won't magically become one. In the words of my (John's) mom, "you'll know in your knower." There is no shame whatsoever in declining an opportunity that you know is not right for you, and keep in mind that just as you are presenting your best version during an interview process, the company is likely presenting theirs; that is, if it's bad now, it'll be worse when they aren't projecting a wonderful image.

Alright, you found the job, applied, made it through the interviews, and have landed yourself a job offer. What happens now? Accepting a job offer can feel almost as nerve-wracking as interviewing. Was the offer high enough? Do the benefits match up with what I was expecting? Can I change the start date? At the point of receiving an offer, a candidate has quite a bit of negotiating power. Even as a fresh grad, it will not hurt to look at your incoming offer and ask for a few thousand dollars more per year, or an additional week of PTO (Paid time off), prior to accepting. Negotiations are expected and are often a part of the process, so don't be too afraid to ask for what you want. Early career salary can set you up for financial success, but not all success is money. Maybe that extra week of PTO is all you need to be happy starting out. The worst they can say is their original offer stands. Be reasonable though – A fresh Civil Engineering graduate won't be able to ask for $250,000 annually to kick off their career.

If you are lucky enough to be faced with multiple offers on the table, analyze your choices based on total compensation, interest in the industry/product, and the general feeling you had during the interview. These categories are weighted differently for everyone. Some find emphasis on making as much as they can as quickly as they can, and others want to have the most relaxed work environment possible. While the choice is ultimately yours, if you plan on staying with a company for an extended period of time (say, five or more years), maximizing your salary at the start is key to keeping pace with the market because often annual raises within a position are in the 3–4% range, and quite often even less. Otherwise, even a low salary at one job can be offset by switching companies after two or so years. I (John) have gotten as much as a 40% gain in base salary from switching jobs, so you can really gain some ground switching when you need to. We'll address how to decide on crossing that bridge when you get to it in a later chapter.

Ultimately, the hard part is over, and you are 100%, without a doubt, a full-fledged engineer. You even have the paycheck with the title on it to prove it! How do you go about making the most of it? Many people, particularly in previous generations of engineers, have had a habit of landing a job out of college and sticking to it until the day they retire, or are laid off or otherwise forced to leave. While this is a completely valid – albeit maybe unlikely – approach to your professional career, this method will not necessarily set you up for developing into the best engineer you can be. To achieve all you want and desire out of your career, you may have to explore the market and put yourself in positions to succeed beyond where your current role is and possibly beyond where your current employer can take you. That may even include changing geographical areas, so we encourage you to be open to considering other things, to the extent you're able to.

We completely understand having areas you just don't want to go to, or that some places won't be feasible for various family or personal reasons. The following chapter will help outline how you can set yourself up for success and maximize your career growth during the formative years.

CHAPTER SUMMARY

Looking back, that's a lot to take in. Acquiring your first job out of college can seem daunting, and even impossible at times, but in the previous chapter we provided key actions you can take to better position yourself in the eyes of potential employers. As you search for opportunities you will find many challenges, from applying to what feels like a million different jobs, to building a resume that properly highlights your skills, to surviving the interview process, to negotiating and accepting your first offer. However, you will also gain key experience that will help you when it is time to advance and grow through your career as you move jobs or industries. That first job is absolutely there for the taking, and the right door will open up for you, but you may have to knock on a lot of figurative doors to find it.

What main things we want you to take away from this chapter are as follows:

- Don't panic if you are nearing graduation without a job offer, it is very common to leave school while still on the hunt for a job
- Utilize the network you built in college to identify any available opportunities
- Master the filter tool in your job search; identifying what you want from a job in terms of role, industry, compensation, and location will assist you in cutting out positions that are a waste of your time
- No position is a lifelong commitment, aim for the job of your dreams but accept the fact that gaining experience can help you attain that dream job further down the road
- Tailor your resume to the job you are applying for
- Hiring managers love numbers. Let us say that again, hiring managers love numbers. Quantify everything you can in your resume

Putting yourself out there for professional scrutiny can be intimidating, but like most things in life, the interview process gets easier with practice. Reach out to your network not only for job leads but also for advice and tips on how they found their way into their career. Most importantly, do not get discouraged as the process plays out. Stay positive as you're searching and don't lose the forest for the trees. You made it this far, and you're not going to give up now.

NOTE

1 Something project managers (what I (John) have been most of my career) do is talk a lot about known unknowns versus unknown unknowns. Put simply, a known unknown is something like "We know we need a machining technology to make this part, but we aren't sure which one", while an unknown unknown is more "we don't even know how to make this".

5 Maximizing Growth During the First Two Years

You got it, the job of your dreams! Well, maybe not the job of your dreams (though we hope that is the case), but a job nonetheless. You are an engineer cutting your teeth in your very first position. The career you have been dreaming of and working ever so hard for has finally come together! Now you just sit back, do what your boss asks of you, collect a few hundred paychecks, and then retire content in the knowledge that you made the most of your opportunities. It's that easy!

Or is it?

While the above is definitely one of the routes you can take, and though the experience gained in university classrooms would suggest that those in positions of power and mentorship, such as professors, advisers, managers, and supervisors, will provide guidance and a path to success, the truth of the matter is unless you are very lucky, your manager has much larger and more pressing concerns than expanding your career. That is your responsibility and yours alone. What can you do to get a head start?

ASK ANYTHING, ABSORB EVERYTHING

As a newly minted engineer or professional, the first few months – or even years depending on your position and/or industry – can feel like you are drinking through a firehose. There will be more information coming at you faster than you can possibly know what to do with fast enough. You have to assimilate to a new environment, learn the protocols and standards for your department and position, find your place among your team, familiarize yourself with any number of software packages and programs, figure out how to transfer your book smarts over into real-life applications and find confidence in yourself to put yourself out there during opportunities when you feel like you have valid input. It truly is an intimidating place to be when you're just starting out, but the quicker you dive in, the more comfortable you will become. As you continue to get engaged and up to speed, you'll see just how much there is still for you to learn, and that's one of the most exciting parts about transitioning your role from a student to an employee.

Among the first processes you'll encounter, onboarding is one of those aspects of a company that you have no way of vetting out until you are already committed. To make it even more of a challenge, this is something that can vary heavily from company to company. Some companies, typically those on the larger scale, have pretty well nailed down the process of integrating new employees into the fold, while others feel like they had no idea you were supposed to be there on day one and everything

DOI: 10.1201/9781003510901-5

is a scramble. One of these situations is clearly superior to the other, but it is possible to make the best and worst of both. And we note, size is not a guarantee of a smooth onboarding experience, so even if you do join a large firm, you still may have to experience a few bumps in the road as you get up to speed.

It is a lot to take in, we know, but every day that goes by and every project that you make progress on will be another step toward gaining confidence in yourself. **Knowledge is your greatest tool in life**, particularly in your professional life. A key factor in maximizing your growth early in your engineering career is maximizing your knowledge base. This does not just apply to adding to your engineering skills, though that is a clear and obvious path. While you are new, ask anything and absorb everything.

What qualifies for asking anything? As an early career engineer, you should be actively seeking information among things such as: The nuances of your position as it relates to the company, how problems you are facing have been tackled by other engineers, steps taken by those further on in your intended career path, steps taken by those in adjacent career paths, general office relations (such as how to interact with the grumpy employee archetype that seems to permeate through every engineering department), what online resources are most useful, and a million other questions that will arise during your day to day job functions. Someone – somewhere – most likely has the answer, or at the minimum, knows where to point you to find the answer. One of my (John's) favorite ways to ask questions in the workplace may be a bit peculiar, but can be a great way to break the ice: "Talk to me like I am a gifted five-year-old." This particular question suggests that you probably have an idea, but need to know the details surrounding it. Details are extremely important! When you are just starting out and at the very dawning of your career, you need to understand a lot of things, including why something is done, and not just how it is done. We encourage you to do what you can to dive in and ask questions whenever and wherever you can, about any topic that is related to your work, but do your homework and come prepared. Much like a job interview, you shouldn't be asking questions you could easily find out on your own with the resources you have available. Put the work in toward learning on your own, and you'll be rewarded with people who care a lot more about the outcome of your initial learning and absorbing what you can. It's a natural human response: We care more about those who care themselves.

You may find questions you ask were directed at the wrong resource or perhaps received a lackluster answer. Part of this action of asking anything to absorb everything is learning who is and who is not a valid source of information. As you progress, you will also learn how to phrase your questions in a manner that can lead to a correct answer sooner, with less confusion. Communicating out, as speaking with your peers across the organization is called, will be an invaluable resource and skill for you to have at your disposal as you advance throughout your career. In engineering, you won't always speak to just engineers, so it's a great practice to start early. As an aside, if you're wondering if you know the engineering side of the process well, just start trying to explain it to a non-engineer within the organization and see where your gaps are. Communication is a key skill in your career, and the more practice you can get, the better. The opposite of communicating out, by the way, is not communicating "in." The corresponding opposite is called "communicating up," and this addresses how you speak to managers and others above you in the organizational chart. By this we mean that you won't change what you say, but you may need to change how you say it. For example, you'll need to tailor your interactions for the audience and what they are trying to get to. If you had 30 seconds in an elevator with the CEO, you're probably going to explain your role or project (or the value you bring) in a much more condensed way than if you were speaking to your functional manager. Learning how to communicate across your peers as well as above and below your position on the org chart is a critical need you'll bring with you during your career, and again, it only gets better with practice.

In the beginning, if you have a technical or company-specific question, go to the next highest rung on the professional organization chart. Someone with a few more years of experience will likely be able to provide answers that are relevant to your situation, whether this be your direct team lead or a more senior engineer on the team. In the event that they do not have the information or any leads on where to find it, you can continue up the ladder to your direct manager or beyond. Enough of these questions and you will likely develop a direct path to the company gurus for whatever sort of subject you may be researching. We find that asking peers for help first can build relationships in the workplace and provide an opportunity to connect with someone who was in your shoes fairly recently. Additionally, the person asking the question may not feel as much pressure to act in a certain way if they are relying on a peer for information as opposed to someone higher up on the org's hierarchy chart. If the question you are asking has no answer, it may even be a better question than you thought it was originally, and is something that is worth pursuing, within reason. What we mean here is if you have a question about a process in the department, for example, but nobody can answer it, then it means that's not a concept clearly understood by anyone and may be an area of significant opportunity to improve. If you can identify problems and improve them, then congratulations! You're well on your way to maximizing the impact you have in your position, especially as you work in the early years.

For questions that are not technical or workplace specific you will find that it isn't quite as easy as going to the most experienced person in the building and taking their word for it like you would with anything technical. There is a nuance to careers that must be recognized and each individual has their own set of priorities and goals, as well as communication style that works best for them. Let this be an opportunity to learn more about your coworkers and their experience as engineers. Your manager will likely be the best source of information on how to ascend the ranks of your current company as they are the ones who typically will recommend you for promotion and know exactly what it takes to get you there. As for expansion between roles, such as moving from Test Engineer to Design Engineer or Quality Engineer to Program Manager or anything in between, go hunting for these individuals at your company and talk over lunch or coffee. Ask them to give you the pros and cons of the role as well as their day-to-day, how they got to their position to begin with, and any number of pressing questions you may have. If you find that your company either doesn't quite have the position you are seeking or maybe you want to find these answers from engineers in different industries you are able to leverage your LinkedIn profile to make introductions. Utilize the network recommendations, search around for open and active profiles in the industry and shoot your shot by messaging them. Chances are if you send an earnest and honest message asking someone how they found their success, you will receive an earnest and honest answer. In the digital age, we can draw from a literal world of knowledge and all you have to do is know how to seek it out. It is your job to sculpt your career in your best interest, so get as much information as you can from those who have been there before you and use it to drive your career in the direction you want it to move.

CREATE CONNECTIONS IN THE WORKPLACE

We touched on connecting to peers earlier in the chapter so you can utilize the relationships to find information and learn. A level of connection is incredibly useful for accomplishing this goal, since people like to help those that they view in a favorable light. We are sure we have all been there: Some jerk in school asks for help on some homework problem, or maybe your neighbor throws trash on your lawn and then has the audacity to ask to borrow your lawn mower. Chances are you don't have any desire to help that jerk, but if a friend, or even an acquaintance with whom you have shared a positive interaction with, asks for the same favor you would provide support with much less thought and consideration. Humans are social beings and being a positive member of the team will greatly improve your career positioning the majority of the time. All skills and results being equal, would an employer rather promote the friendly and personable choice, or the engineer everyone complains about? Again, we are speaking in general terms here and mention again that we have seen all kinds of things play out in our careers, but the point remains: Being liked for reason is a lot more beneficial to your career than being equally disliked for a reason.

Now, don't take this as a call to schmooze your coworkers or put on a fake persona to force them to like you. Both of those ideas will absolutely backfire on you and you will be worse off than when you started. We strongly encourage you to be your most authentic, truest self. Genuine personalities are easy to spot and hard to replicate so it is best to just be professional and be yourself. Remember, you landed your job, so you belong here as you. Your qualifications have gotten you where you are, so you deserve to be here. Be proud that you made it through engineering school and the challenging job hunt. For all that though, be humble, be genuine, and be willing to learn. A point of vital interest here is for you to not fall into the trap of thinking you are above people in your workplace. **Just because you have an engineering degree, you are not more important than anyone else in the organization**. I (John) have seen this all too often; an engineer will be saying something like "that's above your pay grade" to a maintenance worker and will go off on their own smug way (that is an actual quote I heard a senior engineer say once, and it went over like a lead balloon with the maintenance staff). This kind of behavior impacts your career in a few ways, and all of them are negative: You absolutely will not make any friends treating others in the organization as "below" you, you won't get others to help you (which you're going to need, especially as the new person on the block), and the worst of all, it is demeaning to the others in the organization. When it comes to maximizing your career, one of the most helpful abilities you'll have going for you is simply the ability to be liked – for reason. If your coworkers and peers know that you're the kind of person who just gets work done, you're going to tend to be well respected in the organization. We cannot stress enough how important it is for you to work and relate to those in the organization, and we urge you to treat everyone in the workplace with respect. On more than one occasion, non-engineer operators and employees helped me (John) solve extremely complex problems because they brought their detailed knowledge of what we were working on, and we all checked our egos at the door. When you are an employee, you are on a team with the other employees, and teams win and lose together. Regardless of the degree you have, the program you attended, or the perceived value when compared to your peers, you are all on the same team, and each of you has a position to fill.

Teams win together, and teams lose together. The more you can focus on helping move the team forward, the easier it will be to get to a positive outcome. As you continue working and solving and making it less and less about who gets the credit, you'll discover that often teams come into sync easier, faster, and more seamlessly. There will, however, always be people who care entirely too much about who gets the credit, and we encourage you to keep that in perspective. All you can be responsible for in the workplace is you.

Departments vary in size and median age and about a million other demographic factors, so to begin you may not know how to approach your coworkers. Making connections as a traditional student in college is definitely simpler, with the tighter age range and heavy social emphasis a lot of people take on their lives. Meanwhile, your coworkers can be anywhere from their early 20s to mid-60s and beyond (one of Sean's coworkers was into his mid-90s) and many have families that rightfully have the vast majority of their focus outside the workplace. It's going to be up to you to find ways to connect with your colleagues and coworkers, and while we aren't trying to tell you to go make best friends with every person you meet at work, we are saying it can benefit your career to have some people in the workplace whom you know very well, and quite frankly, good coworkers make the work week that much more fun. Forty-plus hours a week for 50 weeks a year is a lot of time to spend with a group of people you patently dislike. There are a lot of ways you can do things to get to know your coworkers, though some companies make it easier on you than others do. Some places do regularly scheduled team-building outings, and others don't, so while these are great, if they are not available to you, or simply aren't your style, there are tons of things you can do on your own, both inside and outside the office. Depending on the rules and expectations of your employer, maybe you could do something like a virtual happy hour with those coworkers who are in a different location than you, or even host a watch party for your favorite sporting event or movie. I (John) have even had the benefit of having a full-fledged movie screen available to us in audio lab, so we did movie and video game nights in there after working hours. Whatever method you choose, there is no substitute for real connections with those around you.

TAKE CHARGE OF YOUR OWN GROWTH

As we look at how you can maximize your growth in a new career, we'd be remiss if we didn't stress over and over that there will be nobody in your career who cares more about your career than you do. Others will care, others will help, and others will absolutely guide you, but none of them will hold as much personal interest in your career as you do. As we mention this, we again reiterate that taking charge of your own growth means you're going to have to grab the proverbial bull by the horns from time to time, and totally immerse yourself into the world around you. As we have said before, you want to make sure that you are happening to the world around you, and not letting it happen to you.

One of my (John) favorite tips for people looking to develop in their careers is to ask them what is the one part of their job that people dislike doing the most because that is likely something worth getting good at doing. What do I mean here? As a figurative example, say you're an engineer and you find yourself designing parts along with people on your team, maybe one of the least popular tasks could be doing all the documentation required for the release of these parts and drawings. That documentation can be a great place for you to shine because it's unpopular for a reason. It could be unpopular due to being a difficult task, so you have an opportunity to develop and demonstrate your mastery with something challenging, or it could be unpopular because it is either an unnecessary or just a poorly defined step (trust us, those happen all the time in virtually every industry, even more frequently than you may think). In that case, you have an opportunity to show you can think critically and analyze processes that need improvements. Unpopular tasks are unpopular for a reason, so if you can identify why nobody likes it and improve upon it, you put yourself in a position to succeed.

Taking this approach will work across a lot of times in your career, whether you are just starting out or if you have been around for a while. Finding out what bothers your peers or what barriers to success the others around you run into can be a huge boost to your own success, if you can make pain points go away. I (John) had a manager once who kept telling me that the best thing I could do in my position was to help someone in another position in the organization. What that means is that you need to consider how your work and position impacts others, even if you don't realize your work will (Spoiler alert: It DOES). Consider your organization as a machine, and now consider your job as a cog in that machine. You have a job to do, so you likely take inputs from someone, and you likely have outputs that go to someone else, even if you don't think you do. A helpful tool you can use to frame your thoughts and develop an understanding of how the pieces fit together is something called a SIPOC[1] diagram. In any case, you need to remember that others will be taking the artifacts you produce and using them for their own jobs – everything in the business world flows down to someone else.

Your job doesn't have to just be improving what is broken and things that people dislike, though that is among the best ways to get involved and make a name for yourself. It's critically important that you work to find your niche in the workplace, knowing that may change over time. If you are just starting out, you probably don't know everything you can even do, much less what you like to do in a professional setting, so we encourage you to take on new responsibilities and try new things at work. If you're a mechanical engineer, for example, maybe you'll learn you really like doing modeling and simulations, only to realize a few years in that you don't want that to be your entire career. That's absolutely okay. Humans grow and change, and what you like doing may change too, but even so, we highly encourage you to spend some time and really dig into what you want to do, as well as try everything you can reasonably get your hands on. In fact, one of the best ways to learn what your organization does is by rotating around different departments, positions, and even business groups. Some employers offer these as really structured opportunities which may be their own early-level positions, often called "leadership development programs" or something similar. In these, you will actually take a period of time, typically between two and four years, and actually rotate through different positions every 4–8 months. These are great, but can be rare and are often quite competitive to land, so if that is not an option to you, you can even see if your management can help facilitate you shadowing other departments or positions for more limited assignments, like a week or two. If you can do this, we recommend that you do it after you've been in the organization at least long enough to start feeling comfortable in asking questions and seeing how the parts of the org fit together.

DIGGING INTO CHANGE

Now you're contributing to the workplace, you're building connections, and maybe most importantly to your career, you're really gaining some experience in just what your role is and what the expectations and responsibilities are from you. As we just mentioned, you may be interested in changing some things around and starting to look at other opportunities within your own company (we'll get to outside opportunities a little later), so you start looking at ways you can get some exposure to other things in your position, or maybe even changing to a new position. We'll pause here and say that we didn't time bound this for a reason – your timeline for learning new things is yours, but we'd guess you'd start actively looking to take on more responsibilities in your role around a year or so in. In any case, how do you go about growing your horizons and finding new things within the workplace? If it were still college, you'd just sign up for another class or take an elective and run with it, but that's not how it works out in the workplace, right? Not necessarily. Sure, you probably can't just volunteer yourself to be on something for 15 weeks like you would with a semester, but depending on the employer, you may actually be able to dig in some and take classes on a particular subject, or particular element of a job.

Depending on companies, positions, and industries, you may be surprised to find there are a lot of formal training options for various topics, usually given in the form of offsite training, or possibly videoconferencing. Check with your employer and see if those are an opportunity for you, and even if they aren't for reasons ranging from business need to cost, you can still express your interest in learning those topics. While you may not be able to attend a formal training, you may find out that another engineer on the team has a project involving that topic and is in need of some help. We won't tell you that you will always get what you want when it comes to training, but (1) you will never get help if you don't ask for it, and (2) your management won't know what you want to get out of your career if you don't tell them. Again, you have to drive your career, and communicating what you want is a huge part of it. In our experience, more often than not, managers are willing to help get you training because it often benefits them and enables you to take on more responsibilities, but there has to be a balance to things. We understand an employer can't just send you to class after class and just train you ad infinitum for you to go work for a competitor, but there is usually some benefit to them as well. We'll dive into this topic more in the following chapter, but for now, remember that learning and growing your career is a dynamic action, and you are in charge.

MASTER THE WORK–LIFE BALANCE

This may be a surprise to you, but being out in the professional workplace is a big change from the normal you knew in college. It seems obvious in hindsight, but the ol' 9-to-5 gig is a big difference from your days of sleeping until noon and only having classes in the afternoon (we obviously are poking fun at the classic college stereotype here, but the point is valid). When you start working full time, things change dramatically; you can't just skip class because it's a pretty day, you'll need to be on a set schedule, and you're saying goodbye to spring breaks (along with summers and having basically all of December off!). This is a change for sure, but it doesn't have to be a bad one. You can certainly overcome the initial difficulty, but you're going to need to be disciplined to do so. You'll need a set schedule, and as a part of that, we really recommend that if you're not good at time management now (i.e., in college) you start taking steps to get there. Something that worked well for me (John) when I was in college was the concept of "working for the weekend," where I tried my best to have all finished all the work I needed to do mostly completed by the weekend, so when the weekend came, I could just enjoy my time with friends. I carried this kind of routine over to earning my first graduate degree, where I worked on it Friday afternoon, then more or less Sunday through Tuesday. While that doesn't mean I was working all day Friday, I did have a routine, and that leads us to our point: You have to have balance in life! Work/life balance is said so much it's nearly become a cliché, but we cannot stress enough to you how critical this will be for you to master in your career and personal life. Sometimes work and school are stressful, not to mention life as a whole, so if you're struggling to budget your time, you're always going to feel like there is a burden over your head. That is absolutely going to wear you down to the point where you're going to feel your personal relationships suffer, and that's a spiral that can get out of control quickly. It's important then to have a plan for keeping some balance in your life, especially as you take on more and more responsibility and progress in your career. There's probably going to be times when you need to work more than you planned, and that's ok, but make sure you set some boundaries in life, especially with your employer. A paycheck is not worth your health deteriorating because all you do is work and stress, but never take time for you. We're big fans of getting in a routine for exercise to help blow off some steam in a healthy way, but that can look different for different people, so we encourage you to find what works for you, and make time for it regularly in your days and weeks. Maybe a walk in the woods is just what you need, or maybe it's painting or playing an instrument or whatever works for you, but make sure that you are allowing yourself time to unwind. When we say a work–life balance, we mean truly balancing work with life AND relaxation.

If you feel like you can't get everything done that you need to, a really good place to start is in evaluating your time management. There certainly are factors that can change how much time you have available, like having to be the caretaker for someone, or otherwise having a special situation, and we absolutely recognize that. In general, though, you'd be surprised how much time you (1) have available to you, and (2) how much time people waste in a given day. When I (John) was working on my graduate degree, I had just started a very demanding new job, started dating my now wife Monica, and was traveling across the country roughly one week a month for work. On top of that, I still made time to go swim, workout, get open-water scuba certified, and enjoy time with friends. There are legitimate ways to not have any additional time, and there are imaginary or self-inflicted ways to not have any time. We want you to make sure you know which category you fall into, as that is going to drive your response in adjusting as needed. If you think you don't have time to better yourself, we invite you to take a look at your time usage, especially in apps and on social media. Sometimes, bettering yourself means spending less time binge-watching your favorite show for the 3rd time, or doing less "doomscrolling" after work. While it can be a change, and even feel like a sacrifice you're making, if you are someone who is actually wasting a lot of your free time (and we've truly all been there), you'll be surprised how much reducing that wasted time and channeling your effort toward something productive can positively impact your mental state and overall health.

Work is good for humans, but working to the point you're burned out and miserable is not, so make sure you're working on setting and maintaining clear boundaries in life, with friends, family, and your employer. You can't be the end-all for every single problem your company faces, no matter how much some people may like you to be. There is virtually no faster way to force an employee to burn out than by treating them as the "fix-it" button for every single problem, and while we're absolutely still encouraging you to take on those challenges and learn the part of the job that nobody likes, we want you to feel empowered to set firm boundaries and push back when you need to in order to protect your own well-being. Whatever brings you balance in your work and personal life, it's critical that you have a routine in place to keep things exactly that: Balanced. There's no time too early to start, so if you're in college or if you're 20 years into your career, find what helps bring order and peace to your work life, and above all, remember to take care of yourself. We have a "grind" culture that says you need to work 24/7 and celebrates "super" entrepreneurs who seemingly never have time to sleep. While we applaud what they do, we want to remind you that usually that amount of work is not sustainable and will take a significant toll on your personal life too. It's never too early to start adjusting your life to have better balance, and if you're reading this, it's never too late either.

To begin, you can make a list of things you want to accomplish and go from there. We'd recommend considering goals on the very short term, like within this month, or even within this week, and expanding out. For example, a week-out goal could be "I want to read 5 days this week," while a month-out goal may be "I want to have read a book this month." What you're going to do with this is list out goals from the very short term to the very long term, and then define the steps you need to take to make this happen. As the saying goes, what gets tracked gets improved, so you'll have some data to help see where you are in and out of balance, as well as what you can do to modify your day-to day to get there. Goals should be reevaluated pretty regularly, so consider this an opportunity to create your own feedback loop. Our recommendation is that you review your goals and where you are at least every double their duration: Weekly goals every two weeks, yearly goals every two years, etc. A common mistake when planning out goals is for people to overestimate what can be done in the short term, and underestimate what they can accomplish in the long term. Sometimes these estimates are way off, and people really miss the point. You've probably heard the "SMART" acronym by now, but if you haven't, it's a great way to set goals. They have to be

> **S**pecific, **M**easurable, **A**chievable, **R**elevant, and **Time**-bound. Try to make goals that meet each of these criteria, and you're much more likely to not frustrate yourself by never achieving your own goals. One more thing we leave you with is to physically write goals out, whether on a piece of paper, or a white board, or even electronically on your phone. Writing things changes how our brains process information, and we want you to achieve your goals. Most importantly though, once you've set a goal, you have to be consistent in your pursuit of that goal. One of my (John's) friends and colleagues from college, John Elam, has a saying: "Consistency always beats intensity.

Set a realistic goal, stick with it, and look back at your progress.

Taking care of yourself can have a lot of different faces from person to person, so what works for you may not work for your coworkers, and we remind you that not every single moment of every single day has to be productive. If you need a day to just sit in a hammock and read a book, then, by all means, take one. If you need a two-hour long gym session on a Friday night, then, by all means, take one. The whole point here is to find the balance *you* need to be well-rested and successful in your endeavors. Balance means not doing things to excess, whether that be work or something else.

IDENTIFYING BAD MANAGERS AND HOW TO MITIGATE THEIR IMPACT

At some point in your career, you're probably going to run into a "bad" manager. The reason we put bad in quotations there is because we want to make a point right up front. Truly "bad" managers are rare. They aren't impossible to stumble across, so we won't say you'll never see one, but there is a difference between a bad manager and one you just don't like. For completion, we'll talk about both here, starting with a truly bad manager.

As mentioned previously, you absolutely do not need to be in a place where your safety, integrity, or moral convictions are jeopardized, so if that's the case where you are, you need to leave sooner than later. That aside, what makes a bad manager? A truly bad manager is going to be one who actively keeps you from succeeding, growing, or improving. This can take a lot of forms, but again, they are quite rare. Among other things in evaluating if you truly have a bad manager, we encourage you to consider a heuristic (if you aren't sure what that is, a heuristic is kind of like a mental SWAG, which we are certain you didn't make to the end of engineering school without knowing the meaning of SWAG) called "Hanlon's razor" in your evaluation. Named for the Pennsylvanian Robert Hanlon, Hanlon's razor says "Never attribute to malice that which is adequately explained by stupidity." To paraphrase, all things equal, it's a lot more likely that someone didn't do something because they didn't know any better than it is because they don't like you. We encourage you to be objective and ask yourself are they actually a bad manager or are they just someone you don't like to work for, which lands them in the second category.

This second category then, must be different from a truly bad manager. We think of the second category as someone who can still hinder your growth, but in this case passively. Hinderances of a passive nature stem from a lot of places, but they can manifest themselves in the form of incompetence, general disinterest, or nonchalant behaviors regarding your career. They aren't actively keeping you from succeeding, but they aren't helping you either, and that's bad for your growth. If you're stuck somewhere with a manager who is either a passively bad manager or one who you don't like working for, you do have some options. The first is obviously leave, but that isn't necessarily the correct option. The next option is to speak to them about what they can do to help you succeed, kind of in the form of "when you do this, then I experience that as a repercussion." This is better than just assuming you need to leave and get out from under every manager you disagree with, but this approach doesn't always work, either from a lack of relationship with your manager, or a lack of accountability culture in your employer, meaning either you aren't comfortable having this discussion, or they aren't held to the standard of helping their team develop, respectively. In either case, you're likely not going to get very far.

Another approach you can take in this scenario is going to involve a little bit of work on your part but provide you with some applicable self-development in the real world – the kind of driving your own career we talk about in this chapter. To do this, you need to ask yourself what makes this manager–employee relationship not click. Is it that they are too busy to manage you, do they assume they are smarter than their employees, does your project or assignment fall on their list of things they don't care about, or are they simply just not up to the task of management? (This happens more in a professional setting than you may think, unfortunately. When you have a moment, look up a phenomenon called the "Peter Principle", coined by Laurence J. Peter. He also authored a book of the same name). In any case, if you find yourself in a situation with a manager that isn't helping your career, one of the ways that's helpful to mitigate that to yourself is a two-step process.

Step one is to identify what they do that is not working for you, and (the more important part) step two is to think about how you would do better. Put yourself in the manager's shoes and come up with your own solution to this problem. Think how you would handle this situation if you were the one running it or in charge, because this does a couple things for you. It firstly changes your perspective, which is always important. You'll find pretty quickly that being someone who is all doom-and-gloom all the time is exceedingly stressful on yourself, and won't make you too many friends in life. You need to be in the perspective that things can be changed for the better, and aren't just a mess forever. Second, it forces you to avoid a dangerous logical trap called an "argument ad hominem," which means you're against the person specifically, not the idea. This is a pitfall because you want to stay above that level of noise, and not bash people. The best solution will attack problems, not people. Finally, and what is my favorite about this, by thinking of a way you could do it better, you take ownership of the problem, even if just in the limited capacity you have in your position. Taking ownership of something – anything – puts you in the driver's seat. We want you to drive your career, whether that is your own development, or in being in a position to fix problems. "Sounds great guys, but I am not a manager" you say, and we hear you. We aren't claiming you can just wave a magic wand and your work troubles disappear, but we want you to focus on what you can control, and do your best to learn from the situations you find yourself, so later on you can do better when you can control things.

There are resources well beyond the scope of this book, and quite frankly, they have done it better than we can, so we direct you to a pair of our favorite books on the subject. Shawn Anchor's *Happiness Advantage* and Clay Scroggins' *How to lead when you're not in charge* are both excellent sources to help you position your responses, and we're sure other works like these are out there for you as well. We can't overstate how important it is to take full ownership of your career, when and where you can, and both of these books will help you identify the areas of your life you do exert control over. Taking control back over your career can help you mitigate the impacts of having a manager who doesn't really care about your development, or worse, a truly bad manager early in your career.

FINDING YOUR NICHE

An important part of growing in your job is doing what we mentioned earlier in this chapter and finding the part of the job that everyone dislikes, but what about the fun part? If you are given some time and resources, you'll likely naturally find your niche at work – that thing you just love doing and always look forward to getting to do. This is both fun and significant, because people naturally tend to be good at things they enjoy doing, and vice versa. You'll automatically enjoy doing something you're good at, and you'll get better at it if you enjoy it. It's a circle that drives itself. Use this to your advantage and harness it. Sure, you're going to need to do things you don't like to do as a key part of your career, and if you want to go far, you're going to need to get good at things nobody likes. But – and we want to stress this point – you can also enjoy your work. Niches can be literal niches, and be the kind of obscure thing that very few other people want or even know how to do, but they can also apply in a more general sense and just be something that you're naturally good at or enjoy. Both are great things to discover in your career, but how do you go about finding one, especially if you're just starting out?

The good news is by the time you make it to the workforce, you already probably have a decent understanding of your preferences. For example, virtually all engineers leave college already having strong opinions on their favorite CAD software or programming language (or if you're like John was, all programming languages were equally bad and he knew he didn't want to work in them). Use things like this to your advantage when you get out into the workplace. Did you discover a really cool trick in a software package you used in college? Find a way to apply it to something you're working on now. Work with your management and senior engineers to see how it can apply to something that's a pain point in the organization. Doesn't look like you can work it in? Get creative. Was there something you stumbled upon in college that really interested you? See how you can learn more about it and work it into your employment. Niches also don't have to be solely work-related things that fit in with what you want. As you're looking to find a balance with your work life, you may notice that you're really good at other things you picked up along the way, and they can turn into your hobbies or side gigs if you're ever interested in doing so (we absolutely encourage you to have hobbies you aren't trying to make money on, and we don't think the goal of everything in your life should be to turn fun things into work). As you work, grow, and develop as a person, you may discover that your niche is way outside of the work you do as your 9-to-5 job, and that's completely okay.

I (John) have known engineers who turned into doctors, winery owners, and even part-time repair technicians just as a result of following what they were most interested in. Like we said earlier, one of the most valuable things you can gain from your engineering degree is the ability to problem solve, especially as the problems you face change. It's also fine if in the first couple years of your career you haven't decided on what your niche is, and you're still looking for what you love. The important thing is that you're looking, and gaining experience which will help you narrow down what you do and don't like. Our best advice for discovering your niche will involve trying everything you can get involved with until something sparks, and then let your own creative ambition take it from there. Don't be too shy to try things, either inside or outside of work, and ask all the questions you can. If it's taking you a while to get there, don't get too discouraged. Think back to our ship's sails analogy we used earlier on, and apply it to this as well. You'll get there, it'll just take time and consistency.

HOW TO FIND MORE – INSIDE AND OUTSIDE YOUR ORGANIZATION

What happens when you realize that you miss the way school was? Don't laugh. As crazy as that may sound, you will eventually miss some element of school, despite how stressful and busy engineering school seems to be for everyone who goes through it. Whether it is the structured learning, the variation in getting something new every four months, the flexibility of your day, or just the time you spent with your friends, you're more than likely going to miss college, or at least some elements of college. One day you're going to find yourself wanting more, and whether that is professional, personal, or even academic, there's a few things you can do to help yourself out.

As with a lot of things, the first step is for you to uncover what you feel is missing. Do you feel like you're not growing at work, do you feel like you don't know anyone in a new town/find it harder to meet people once you're not in college (both of those are quite common!), or is it still something else? Work, for better or worse, provides us with people we will spend most of our time with, at least during the week. It's awesome if you have some coworkers with whom you actually enjoy being around and become friends with, but if you're really not that interested in spending more than 40 hours a week with someone, we get that too. In my case (John), I've had work friends that I made and talked to nearly daily for years after I left the company, as in the case of Sean and others, and on the other extreme I've had people I worked with that I outright wouldn't talk to if we bumped into each other on the street, so trust us when we say we aren't telling you to try to find all your friends as work. What working somewhere does have going for it as a resource in this case is professional groups, either internal or external. These have names like professional groups, resource groups, social societies, etc., but they are all really similar despite their naming differences. What these are is a way for you to grow your network (remember when you were just starting to network for a job?) in a semi-structured sort of way, and that can be a great source of meeting new people. These, plus professional societies, conferences, and seminars, can be an excellent way for you to build that network around people who have similar working experiences to where you are currently, but they aren't going to help with everything you may be missing from your college days.

What happens when what's missing is the sense of structured learning you had while you were in college? Let's assume for the sake of this that you're in a position where you don't really feel like you're learning new things or even growing anymore, and it's not something that you are going to be able to fix by taking on new responsibilities at work, at least not for a little while. You don't want to change jobs, and you don't want to do something like an additional degree, so where does that leave you? Again, work *can* help. Depending on your employer, they may have some surprisingly good learning resources, like LinkedIn learning or something of the sort, where you may be able to take some classes on things that aren't even germane to the core business of the company. When I (John) was at the Bay Area tech company I'll speak to later, they had unlimited LinkedIn learning courses for ANY topic that was available on the site. They also had a massive engineering library full of resources and textbooks on nearly any subject, but I understand this is the exception and by far is not the rule. Where do you turn when that is not the case? Again, technology is your friend here, and there's tons of places to turn to when you're looking to get going on the learning front. Sources like YouTube, AI-based models, and even traditional academic institutions are fantastic sources of information, and mostly they are usually free. In some cases, like the last one I mentioned, the universities even publish short, condensed forms of their entire classes for free, and this can be a great place to kick start your learning. Something I (John) do frequently when I mentor people is point them toward the power of AI-based learning, like Chat-GPT and other sources, which are both growing in number and capability all the time.

We'll pause here to say while there is enormous power in AI, there are also some areas we'd advise you to have a cautious eye on what it's telling you. For example, if you just ask an AI model to write you 1000s of lines of code, you're probably going to have some issues and it won't all work exactly as you intended it. These models are growing every year and becoming more and more accurate, but they are still somewhat prone to making things up or giving false information. One of the things we use it for quite a bit is to ask AI to explain things that are extremely well-known and documented, but in a way that is different than a textbook may explain the material. I have asked AI to explain more complex theories to me, for example, the Navier–Stokes equation and what it actually tells us in regard to aerospace engineering. If you want to do some structured learning outside of work but aren't wanting to do additional degrees, try using AI to build you a study plan. Prompt it with as much information as you can, like what you're wanting to learn, where you're starting from, and what your level of experience in related fields is. Give it a try and see what you learn. We'll also say there is a lot more to be written on the subject of AI, and plenty already has been written by others far more knowledgeable in the subject than Sean and I are. As you've no doubt learned in getting your degree or working to this point, technology can be an extremely powerful ally, IF you use it well, such as for structured learning.

It won't be our first recommendation just because it is a huge commitment, but if what you're missing from college *is* college itself, then it may be a time to consider going back to school, for either a graduate degree, a certificate, or a particular program or even a second undergraduate degree. Sometimes what you want is the actual, structured learning that you had when you were in college the first time, and that's one hundred percent okay too. We'd be hypocrites if we told you to not do grad school. Having said that, why is it not going to be our first recommendation? Because there is a lot involved, not just monetarily, but also in your time. It's a large cost, and if for any reason you don't finish it, it's a waste of money (on paper anyway). That doesn't mean it's the wrong answer for you; for us and many others, it was certainly the right answer. Just understand what you are wanting and why you feel like you are wanting more, and use that knowledge to drive your decision. It's also okay for you to pursue additional education that isn't related to your career as an engineer (like an electrical engineering going back and earning a history degree instead of a graduate degree in electrical engineering). If that's the case, make sure the cost is worth it to you, and treat it like a hobby. It's not a problem to have hobbies – even expensive ones – as long as you plan it in and treat it like what it is. Don't go down the path of trying to explain to yourself that your hobby is a great investment because you want it to be one. A hobby can be a hobby, and that's completely fine. Much the same as managing your finances though, you have to know what in your portfolio is an asset, and what is a liability. It's okay to have both, as long as you know which category it belongs to, and you are not trying to convince yourself it belongs in the other category.

We think formal education is great, as long as it fits in how you want it to, and can be done without unnecessary debt loading. Formal education doesn't have to be only degrees and certifications either. You can take formalized education in everything from dance to cooking to photography, and that level of structured learning may be just what you need to satisfy the itch you have from college. There's also, generally, a lot of opportunities you can have outside of full-fledged four-year universities like you may have graduated from in the form of local junior and community colleges. These are typically available to some extent in most metro areas, and offer everything from complete certificates to just the option to take classes ad hoc, in areas of interest ranging from welding to cooking to automotive repair to literary history. If what you're craving is formal education, there's likely several avenues you have to fulfill that interest. As in the case of when you were looking at schools the first time, the financial and ease of use advantages of local two-year institutions is hard to overstate.

Okay, education is great, but what about in the case where you're wanting things technology and formal learning can't help with? Maybe what you miss from college is not the structured learning but the social aspect, and if that's the case, we understand completely. Humans are meant to be social creatures, and even if you're more toward the introverted side of the scale, you're still going to want to interact with others at some point. Work can be tough because you're still learning what your boundaries are, and you're to some extent going to be the "new guy" (or gal) wherever you end up, as well as possibly being new to the area adding to the uncertainty. Here, we point out a few things that can help you make sense of it and maybe give you some direction.

Broadly, we'll call it three categories (which can all absolutely overlap with one another): Things that help you grow, things that are just fun, and things that help others. Once you're done with school and settled into the routine of work, you're probably going to be surprised as to how much time you actually have after engineering school, especially if you were working through college or on design teams, or athletics/other extracurriculars. Once you survive the transition and comfortably shift to those regular, steady 8–10 hours a day of work, you'll find yourself with ample time for other pursuits, and hopefully, some extent of disposable income. You can use this to pursue hobbies you have, and a lot of hobbies have social clubs for all levels of interest. Hobbies can help you grow and be just something you enjoy at the same time, and we have encouraged a lot of our mentees to join various social clubs, ranging from classic improvement opportunities like Toastmasters to the local racecar club downtown. There are plenty of social opportunities for you to use the skills you gained in engineering school too, as in the case of Engineers without Boarders or others. That of course leads us to our next point, and before we get there, I (John) want to quote my old manager (the same one we've quoted before in this book) and tell you "It's hard to be helpful and unhappy at the same time."

Never underestimate how much power there is in helping people, either of the impact it can make in their lives, or of the impact it can make in your own life. We encourage you to give back to your community in the way of your time and resources, and the community of Earth as a whole (meaning other places, people, animals, etc. Whatever you're into supporting). If you're missing some social aspects of your life, there are myriad things you can give back to, from mentoring/tutoring at your local junior college to serving at a shelter to helping coach the same sports that gave you so much when you were younger. Making the transition to employed college graduate leaves a lot of people feeling like some part of them is missing or left behind with their old self, and it doesn't have to be this way. The reason it seems something is missing is because once you become a "full adult," things have a natural tendency to become more about "me, myself, and I," and a lot less about those around you. Take a long moment of self-reflection and see the areas you can develop where you feel you're missing something and get out in your community. To go full circle for this section, your employer probably does a lot of community outreach in the form of blood drives, community clean ups, and other sponsored events, and that can be a great place to get back to giving back if you're not sure how to start. Local charities always need help, and this can take the form of everything from working in a food bank to giving back to that sport you always loved as a kid by working as a volunteer coach or referee. There are so many ways you can get involved in your community we're only scratching the surface here. No matter what you're interested in doing, we strongly encourage you do get creative and give back to your community. We're all made better by those who give back. In our opinion as authors, if you're feeling like you're missing something now that you're a professional, you should start with giving back. That doesn't mean you can't or shouldn't identify what you feel is missing, but in giving back, you benefit people other than yourself. In giving back, you can very easily make an impact on the community around you, and that on its own is worth doing. As you proceed with giving back to the community, you'll surround yourself with more people who can help you identify exactly what you think is missing, and what actions you can take to fix it. Giving back in this way is one of the few times that truly everyone wins.

HOW EARLY IS TOO EARLY TO MOVE ON?

This is an entry point of sorts into our next chapter, so we'll introduce it now and dig deep in that chapter. Entry-level jobs are not forever, so how will you know when it is time to move on, or perhaps when it's too early to move on. To us, we say you should generally be in your first role no less than a year before trying to move on to something else, but there's always exceptions. A few exceptions to this one-year rule of thumb would be if you are in any sort of situation where your employer is conducting unsafe, unethical, or outright dangerous actions, including any sort of harassment or persecution. It may seem obvious, but this is wrong, and your safety, as well as the safety of others, is extremely important and can't be overlooked. If you think you're experiencing any sort of harassment, or witnessing unsafe or unethical practices, we encourage you to know your rights as an employee and point you to the US Equal Employment Opportunity Commission[2] as a great first resource to begin your search. From there you can determine the appropriate next course of action and how to best pursue it.

Another reason you may leave before you've hit a year is going to be a case-by-case thing, and that would be if you have an opportunity that you just can't pass up. If you've always wanted to be an F1 engineer and your favorite race team just happened to give you a call, then the answer here is pretty obvious. In my (John's) experience, the moving on has come pretty naturally, and I just knew it was time to move to the next stone on my path. In general, you'll be in your first role for only 2–3 years anyway, so don't be surprised when you feel the need to move on. My first manager after college told me once that if I was still in the same role in three years, then he had failed me as a manager. While not true for every case, you should expect that timeline to be more or less the norm. We tend to say that the absolute shortest length of time you should be in your first role, again assuming you aren't involved in anything dangerous or unethical would be about six months, and ideally around a year before you considered leaving your first role. That doesn't always happen for a range of reasons, so you may be trying to leave before then. There may also be some questions about why you're leaving as you try to transition, so be prepared. Some people (and companies) see jumping ship early or often as a huge no-no, and others aren't that concerned by it, so there's no blanket statement we can tell you to work for every case.

Our advice here is to be honest to the extent you need to be, but bashing your previous employers is usually not going to help your case. Even if you do have some very serious and legitimate claims against them, there's no reason to air these grievances to an unrelated third-party and hurt your own credibility. If you find yourself having to defend leaving "too soon," make sure you know your own reasons for doing so, and be ready to speak to them. In my (John's) experience, there hasn't been a time when I didn't have a good reason to explain away some concerns, and I've just crossed those bridges when I needed to. It's been my experience that when it comes up, other companies haven't pressed into my answers too much, so just make sure that it's a good reason and you can explain why.

CHAPTER SUMMARY

Maximizing your career growth in the critical first few years of your career is a great way for you to continue the momentum you had from finishing college up, but there's a unique and new set of challenges here. All the strategies and tips we presented in this chapter can help you take control of your career and make sure it's working for you.

- Be a sponge in this part of your career – Ask all the questions you have, and absorb all that you can
- You work with other people, so don't miss the opportunity to connect with them
- Nobody will care more about your career than you will, so it's up to you to own you career and drive it
- Changes will happen, and they can help push you forward if you'll let them
- Working is different than being a student, and it's okay if you experience some transitional growing pains as you get adjusted. Remember though, you must create balance in all areas of your life or you'll burn out fast
- Bad managers can make for bad experiences, but they don't have to define you. Keep an eye out for bad managers and what you can do to help them help you in your career
- If you can find your niche, it allows you the opportunity to really drive your career how you want it to go. Don't feel bad if you need some time to do it, but find what inspires you and run with it
- There's a lot more to your organization than just work. Use your position and your community to help fill any voids you may feel as you transition from the college life
- Knowing when to move on is a critical skill for you to develop in your career

Take all of this growth and knowledge with you as you move into the next phase of your career, and leave being an entry-level new grad behind you.

NOTES

1 Suppliers Inputs Processes Outputs Customers, or SIPOC, is a way to list out everyone involved in a particular process.
2 www.eeoc.gov

6 Beyond Entry Level
How to Know and Prove to Employers You Are at the Next Level

You've mastered all the tips and tricks from the previous chapter, and you've now gotten some good experience under your belt. So much so in fact, you're now starting to wonder if you're still an entry-level employee. While "entry-level" can mean a lot of things in today's job market, in general, entry-level covers you from zero experience up to about two years on the job. That's the point where you're probably going to have the skills and experience necessary to start making some upward moves in your career (and it's ok if that point comes sooner or later. We always want to remind you that this is your career, and it won't follow the timeline for someone else.)

Two years is the line we chose here because it's enough for you to get your feet wet and seriously start learning about what you're doing. Depending on your industry though, two years could still barely be enough to even start knowing what is going on in your role. A little segue here: For the sake of simple math, there are about 2,000 working hours in a work year (50 weeks times 40 hours a week). Generally speaking, to be considered an expert in something, you need at least 10,000 hours of practice. That works out to around five years of full-time experience, which is why there's so many jobs that jump up in responsibility and title at the five-year point.

More than just hours worked though, moving out of an entry-level position is about the level of autonomy you now bring to your career. You should need less help than you did (read: It's still ok to need help!) when you were just starting out, and you should be taking on more and more responsibilities in your position. With those first years of working experience, you've started learning how to apply what you went to school for in the first place, and now you're really getting going smoothly, and even wondering what comes next in your career. This can be a lot to take in, and if you're like us, you may still catch yourself asking where your first two years or so even went, after all that wishing and dreaming back in college to be where you are now, and that's completely okay. Life goes fast and most things don't last forever, whether that's good times, bad times, or first jobs. The huge advantage you have this time around is in our experience, it was a lot easier to get jobs #2 and beyond than it was to land job #1. Not only do you have some experience on paper, you have also probably gained some intangible experience, from now knowing more of what you do and don't like, to having dealt with a lot of situations that college never prepared you for. This will help you whether you're looking to change companies, or just take on more responsibilities in your current role or employer. In this book, we've talked about "jumping ship," so to speak, but we don't necessarily think you need to change companies all the time (and in a lot of cases, you shouldn't) to find bigger and better

DOI: 10.1201/9781003510901-6

things in your career, but sometimes you'll need to, whether you're entry-level or not. If you have enough that you like to keep you where you are, that's awesome. Just be aware of how your salary is playing out, because in our experience, after about 3 years or so, your wages will have stagnated some and you can be behind the market rate.

Even that doesn't automatically warrant a move to another company, but you're a lot less likely to pull down double-digit percentage raises in the same company than you are looking to move outside, even for moving up to a new pay band. We're not telling you it'll never happen, but we are saying it's not the norm for people. In our experience, getting anything over 5% – much less 8% or 9% – at once from the same employer is pretty rare, unless you're really low in your pay band to start with. We want to give you what you need to be prepared in your career, so now is a great time to mention the concept of what's called a "compa-ratio," and why you should know about it.

The compa-ratio is a way to show where you are in your pay band, and a lot of employers use it to determine how much room you have to grow within that pay band. Here's an example: Let's say for the sake of the easy math we all love, the true middle (as we're all engineers here, the arithmetic mean) of your pay band is $100,000 annually. You make $90,000 annually, so you're sitting at 90% of the mean, or in compa-ratio terms, you're at 0.9 of the mean. If you made $110,000, then your ratio would be 1.1, and so on. In general, the true market rate of a position is somewhere between 0.8 and 1.2 of the middle of the pay band, adjusted for location, special skills, etc. If your compa-ratio was sitting at say 0.6, there could be a few reasons for that: You're either super new, a poorly performing member of the team, or simply that you're unfairly paid, and you need to make more money. On the other side of that equation, if your compa-ratio is something like 1.6, then you're virtually maxed out in your pay band and you're going to have a hard time pulling more money without a promotion. We mention compa-ratio mostly so you're aware of one of the methods (though certainly not the only one) companies use to determine salaries and raises because we were well into our careers before we knew about them and how they functioned. If you're a top performer, as indicated by your performance reviews and your peers, then you can use compa-ratio to further leverage your argument for more money when it comes up. When it comes up, and at some point in your career it will, it's in your best interest to have as many factual, logical points to use as you can to get not only the salary you want, but the one you can prove you deserve.

With that, money aside, you think that you're ready to move on from where you started though, whether internally or externally, so what comes next? To begin with, and this may seem like stating the obvious, you're going to need to know you're ready to move on. Maybe you just know you're ready, or maybe you're afraid you won't know when the time comes, and it won't be obvious. In our experience, it's usually pretty obvious. You're going to hit a point when you will no longer feel like you are where you should be, either from management, the environment, wanting to do more, or what drove me (John) the most, feeling like you were no longer growing in your career. Whatever the reason, when it's time to move, you'll likely just know. That could be the two years point like we said, or it could be much longer or much shorter, but you'll know.

Once you know you're ready to move on, the next obvious-but-hard-to-answer question is where are you headed from here? Put differently, "I know I'm ready for more, but more *what exactly?*" This can look like all sorts of things to different people, so you'll need to understand what motivates you. For me (John), I started looking to leave my first job because I didn't feel like I was growing where I was, and I was starting to worry if I didn't leave soon, I'd be behind in my career – like some future employer would look at me and go "you've been out of school for *how long?* Why don't you know how to do this?" and that was horrifying. In my case, I knew I wanted to look at external positions and find myself a new company, but again, moving on doesn't have to be moving outside the company you're at now. A lot of this depends on the company you're at and how much room they have mobility there. We'll address how this may look in both cases, internal and external.

Let's start with moving to a different internal position. Even internally, there are a few ways this can go: You could move to a new pay band in the same group, say from an Engineer I to an Engineer II, you could move to a new group as the same pay band, or you could change both groups and pay bands entirely. If it is the case where you'll move from an Engineer I to an Engineer II in the same group, you more than likely won't have to do much, if anything to get that promotion. It's pretty common to naturally come up in the course of your regular reviews and check ins with your management that you are now performing at the higher level, and the promotion will follow. Of course, you are encouraged to have these kinds of growth conversations with your management, and ask questions about your career path and how you get from one place to the next.

Hopefully your company has a well-defined definition of the career levels, complete with responsibilities and levels of proficiencies. If they do, you can use this as a rubric of sorts to self-evaluate and recognize your areas for improvement, as well as take credit for things you already do very well. We encourage you to do this on your own and not just wait for your management to do it for you because as we've said over and over, this is your career. Nobody will ever care about your career more than you, so you need to be on top of what it takes for you to continue progressing. This well-defined rubric is great, but to be honest, only a couple of companies we've worked in had the professional levels this cut and dried, so you'll probably have to do some of your own digging. In this case, you can always use your management as well as the more senior people on your team/that you work with, and start evaluating what they do that you don't. There is also always the possibility that favoritism is alive and well in your employer, and you're just not "popular" enough to get promoted when you're obviously ready for it, and even if there is a position open for you. Another thing we've seen a fair number of times is the "dangling carrot in the future" idea, where they constantly promise you things that don't exist yet. This can take numerous forms, but the gist is usually the same, some form of: "you're well suited to this position/project/posting that won't be an option for a long time (if ever)."

Depending on your personal factors, you may be perfectly okay waiting on the promise of something new in the future, and you may not be. There is absolutely nothing wrong with wanting to gain some tenure and longevity at a company, but we want to again set expectations for you. This is not the same work environment our parents and those before them had, and we want you to be able to make an informed decision.

Staying your whole career at one place and retiring with a pension and a gold watch is now a very rare exception, and is in no way the rule. Some companies and even industries are more volatile than others, so employee tenure and career longevity can vary wildly from place to place. You also need to consider things nobody ever planned to happen, as in the case of 2020 and COVID changing plans for literally everyone on the planet. Keep all of this in mind, and don't take things personally, but if your management is clearly favoring others or dragging their feet, you can certainly factor this into your decisions to stay or go elsewhere when the time comes. We by no means intend this to scare you, but rather point out that life is a dynamic, changing event and you'll need to be prepared for some changes, even those you didn't know you'd make.

An internal promotion can take the form of a natural progression of your role, as we said before, or it can be the result of a more focused effort on your part. While you can't really rush your career development, you can be proactive, taking on additional responsibilities and learning how to progress your career. You may hire in as say, an Engineer I, and somewhere within the next couple years your manager could decide you're ready to go to Engineer II and it happens basically automatically, or you may have to wait for the company to open a requisition so you can apply to the next position where you will have to interview or otherwise prove you're ready to take that step.

Neither of these is necessarily better than the other, and we haven't seen one be significantly more common than the other. It is also common that a department will also have maintain whatever budgetary restrictions it has and could have a true limit to how many employees may be promoted in any one fiscal year. It is unfortunate, but you should be aware that not everything is completely fair, and some people will be in higher levels than you before you think they should be. This can be due to some combination of education, experience, responsibility, and unfortunately, just how likable they are. It is also worth noting that transitioning from Engineer I to II or II to III should be a smooth and expected shift, but transitioning groups or job responsibilities (say, going from Test Engineer to Design Engineer) is more than likely not going to be as streamlined. Making this transition should come with the expectation that you'll need to prove yourself to your potential new manager and/or team by way of some form of an interview, formal or otherwise.

In either case, if you end up being the one to lead the conversation regarding your promotion – we recommend that you be the one to lead this conversation when and where possible by the way – it is highly advisable to come prepared. When it's time for you to speak with your manager about where you think you should be as opposed to where you are currently, walking up and saying "Hey, I think I should be an Engineer II because the title is better and I want more money" is not going to get you there. What any management team needs to see is that you are aware of your contributions to the company and the value you add to the team. You can do this by either taking your company's most recent job posting for the position and providing evidence of where you meet or exceed their qualifications, by reviewing your past projects and listing out all of the times you performed above and beyond expectations for your role, or when you showed moments of growth and initiative. Sometimes all of them together, and bonus points again if your employer has a defined rubric you can use to grade yourself. By providing tangible evidence of your growth in your role it will be easier for your management to run your promotion through their management, and will let them know what could be lost if they don't do their best to retain your skills.

When making a case for yourself, the best building blocks you can use will capture moments where you provided valuable input to a design or procedure, reduced cost on a program, found a way to deliver earlier than expected, mastered a technology or program frequently used by the department, or made the extra effort to provide support for the team. Obviously, your good highlights, right? What about your bad ones? Those can absolutely be used too, when the timing is right. You can cite lessons learned or significant failures, as long as you can prove that you not only learned from it, but that you learned *because* of it. Anything that you have done or learned that helped the department run smoother, faster, or more efficiently is going to help build your case in convincing your management that you are more than just a junior engineer fresh out of school. There is both good and less good (it's not really bad per se, it just depends on how you approach it) in going through the internal application and promotion process. The good is pretty obvious, in that you're already there, and the people you're interviewing with likely have at least some level of knowledge surrounding your performance so you're not completely starting from scratch. The less good part is that they will also potentially be aware of your shortcomings within the company too, but like we said, it's really an opportunity if you approach it as one. Anything you didn't do well is a great opportunity to explain how you've learned from that experience, and why you're a good fit for this. The other thing that can be less than good is that your current employer can do some weird stuff when you interview. I (John), for example, once had to do an interview for a position at my manager's level *with* my manager in the room. Why was he there? I have no idea. It wasn't a secret that I had applied to this position, but he wasn't even on the meeting invite, so seeing him in there was a kind of odd experience. Again, it was not bad per se, just different from what I was expecting (by the way, I did not get that job. Later on, I was glad that door closed for me too, and not getting it ended up being better for my career anyway). The only other "negative" we can think of in the internal interview process is more of a perception thing, and that is whether you get the promotion or not, you still have to work with these people, so don't burn any old bridges, bomb the interview, or say anything you'd regret later. When you are asked "why are you trying to leave?," it's always best to shift and keep the focus on the future, not the past. A great answer is something like "I'm ready to take on more responsibilities/hone my skills/continue learning" or something you can spin along equal lines, and a bad answer is anything that may sound like "my current group is terrible." Even if it is true, that's a statement that won't help you in any interview, least of all an internal one. Responding to those kinds of questions by focusing on where you're headed is a great strategy for both internal and external interviews.

Great, that's the gist of the internal promotion process, but what if you are looking for growth by transitioning to a new company? Whatever led you to look outside, whether by choice or necessity, you will have to prove to the hiring managers not just that you successfully completed engineering school but have also now spent the last few years gaining real-world experience that benefits them. This is essentially what you'll do when fighting for internal advancement, only the communication is different, as we alluded to earlier. What do we mean by that? Well, when providing examples of career experience to your current company there is a deeper level of understanding of both the importance of the projects you touched, as well as the impact you had (obviously, since they are their projects too!) So, on the positive side, when speaking internally you are able to provide more detailed examples of exactly what you were doing without worries of Intellectual Property (IP) being shared or things being lost in translation. However, the advantage of an external opportunity is the lack of preconceived notions and biases as well as a potential variance in what they deem important. Sometimes you just do not get along with a manager, or maybe your extra effort is downplayed by your department because of poor work culture, or perhaps your best skillset is being underutilized and not taken into account when reviewing your career with the company. What this means is you can now frame your experience and skills in a way that works best for you and frames your qualifications and experience in a way that you want, without being burdened by the preconceived notion someone may have. The tradeoff boils down to *a priori* knowledge versus *a posteriori* knowledge to borrow Rene Descartes' idea. In short, your current team will know more about your successes, failures, skillsets, and projects, but an external group will be looking at a blank canvas, so you have more opportunities to start fresh. As you look internally versus externally, you'll have to balance knowledge against opinions.

Alright, which move is better? Internal or external? Well, just as with every other career decision, that is entirely up to you and your situation. While we cannot tell you every step you will take in your career, what we can do is break down the choices and let you make informed decisions. As mentioned previously, an internal move is the most straightforward option that will typically happen on its own accord (mainly applicable to the early stages of your career and is entirely dependent on how your company handles its department) and you will benefit or suffer from the familiarity and biases of your management. Additionally, there is a higher level of transparency when it comes to what responsibilities and additional tasks you may be taking on while moving into a new role. It is easy to ask coworkers who have previously or currently held the position you are striving for what additional expectations were placed on them as well as what may have been a surprise for them upon taking the role. On the other hand, external roles are a fresh start but with that comes all the uncertainty and unfamiliarity that you have more than likely worked through at your current position. What it comes down to is (1) how badly you want a change and (2) how well are you being treated.

There are numerous reasons to go externally, just as there are reasons to stay. We'll make some general statements to this end, and hopefully give you some leverage to consider in your decision. If you stay internal to the company, just in a new position, you likely gain the benefit (potentially, assuming you like where you're at) of not having to move. You also gain more experience in what you've already started, as well as potentially gain more benefits, as it is quite common for employers to give benefits such as increased PTO, profit sharing, or additional bonuses and retirement contributions to those who have been there longer. If you stay though, you should remember a few things as well, among them that staying longer is not necessarily a guarantee that your commitment to a company will be rewarded in your longevity there. People can stay for decades at one employer, and still get cut just as fast when the chips are down and money is tight. There can also be quite a bit of wage stagnation, especially in your early career if you just stay with one employer. This can open you up to the potential of significant lost wages, and missing out on other opportunities, so you'll need to carefully consider all sides when you are looking to change positions, either internally or externally.

Another thing that may drive you to leave and go external – and if you fall into this category, you're by no means alone – is that you simply outgrew your organization. I (John) have left roles because I was no longer growing as a professional, or as a person. Recognizing this takes some practice and experience, but like everything else we mention to you, it gets easier with time and practice. You'll know you're not growing as you want to once you hit this point, and once you hit it, it can be hard to overcome mentally without a fresh start. Outgrowing your organization can take a lot of forms, including a lack of growth as I mentioned, or it can be a change in management structure, or simply that your personal and career goals no longer align with those of the company. All of these are valid, and can be good reasons to leave. We encourage you to ask yourself some questions before you leave though, because as the old cliché goes, the grass is not always greener on the other side. Just because you want to leave your employer for any of these reasons, it isn't a guarantee that the environment is going to be better for you on the outside.

To that end, we know you can't know everything before taking a role somewhere else, so all you can do is discover as much as you can before going somewhere new. Thankfully, this is easier now in the days of the internet and social media, so you can find out all kinds of things before you join a new company. Take things with ample discretion though, as places like Reddit and Glassdoor are disproportionally filled with people who are all too willing to find the fault in things, and that may not be the most accurate representation of the employer. One of the ways we recommend trying to find out as much as you can is to use sites like LinkedIn to find former employees with similar titles or positions to those you're interested in at a particular company, and ask them some questions. Again, you may be surprised at people's willingness to share information with those they don't know, when given a good opportunity to do so. If you go that route, don't "lead the witness" with them and open the conversation with "I saw online that Company X is terrible, and I was wondering if that's true?", but rather something along the lines of "I am looking at Y position in Company X, and I see you used to have a similar role there. Do you mind telling me why you left?" Again, people are often willing to share their knowledge, they just need a good prompt to do so. Use the resources available to you to make the best decisions you can, and avoid the pitfall of thinking that everything is better somewhere else. We want to help you make sound, data-driven decisions, to position yourself where you want to be in your career.

Whether you are looking internally or externally to make changes, this is a normal part of your career, and it doesn't have to be a traumatic experience for you. It is also something you are going to need to get used to because as an adult, you're going to come across a lot of Crossroads in your life, and dealing with those specifically is the subject of our following chapter.

CHAPTER SUMMARY

Chances are, you're only going to be considered an "entry level" employee for about 2 years, more or less, then you'll likely want to move on from your first job. Even if you stay with the company you started, this is around the point where you're going to want to make some adjustments to make sure your career is going where you intend it to, and that you're not falling behind on your compensation when compared to your peers.

The main things we want you to come away knowing from this chapter are:

- Nothing is permanent, including your first job
- Zero to two years of experience is about the duration we'd expect you to be considered "entry-level"
- Whether you stay internal to your first job or look to move externally, you're probably going to have to interview again, so look for ways to relay what you've done and translate that to what you're looking to move to
- If you're ready to look outside your organization, know what you are looking for and have an idea of what you're trying to get out of it, so you're not just leaving for the sake of it.

7 Crossroads

When you were in middle school, you probably couldn't wait to be in high school. Once you were there, you realized being a freshman was substantially worse than being in middle school, and you couldn't wait to be a senior. Once you were a senior, you probably had a hearty case of "senioritis" – where you absolutely cannot wait to be done with school – until you moved on to your next step, whether that was directly on to college, or into the military, or perhaps a vocation or maybe even some form of service organization. There are a lot of ways your life can go, and that can either be exciting or it can be terrifying, depending on your outlook. You may think of it as an end zone that is always moving; a quixotic chase where you'll never actually reach the end, or it simply may have never really occurred to you before. As we humans grow, the goals we have change, and that's completely normal. As you saw in the previous chapter, first jobs aren't last jobs, and you're going to move on from them at some point. What happens when you aren't just looking to move up in your career, but maybe move to something different entirely?

I (John) knew I wanted to be an aerospace engineer from the time I was eight years old, but that degree still remains to be earned, despite having earned two other engineering degrees. What happened? Somewhere between being eight and being in engineering school, I hit a crossroad, like many of you will (or already have), and I had to pivot. Making a pivot is not inherently bad, and we want to make sure that point is stated. You learn and you grow as you move forward in life, and as a part of that, you sometimes find yourself doing things differently than you had planned. In my case, I started my academic career as an aerospace engineering student, but found that I was much better suited to other specialties in engineering, and that's been what has led to me having the extremely unique experiences I've had. While I wouldn't change anything about what I've done, the aerospace engineering discipline is still what gets me excited, and is the subject of most of my reading. Still, I have yet to earn a degree in that (I say yet, because as of this writing, it is something I am pursuing as a bit of a "hobby degree" and intend to complete at the graduate level, having started a Master's program). Maybe your path looks nearly exactly how you planned it out when you were eight, but I'd guess no matter how close you got, it's not *exactly* how you planned. I certainly own fewer airplanes than the eight-year-old version of me thought I would own by now. The good news in this is you've already encountered your own crossroads, and you've had to determine how to proceed.

These crossroads can come professionally, academically, and even personally, and all of those are okay. Hitting a crossroad doesn't make you a failure, or an impostor (which we'll get to in the following chapter), or even that your plan was bad to start with. Sometimes you pivot because you needed to with what you knew at the time, as was the case with me and aerospace engineering. When I started it in my very early twenties, I just wasn't ready to do it, and I had to wait another ten years for me to learn enough about myself to feel like it was the right time to try it again. You may

DOI: 10.1201/9781003510901-7

see in your case that what you thought you wanted to do was not what you hoped it would be, or you may have stumbled upon things you didn't even plan on and you're considering making a major move. Crossroads and pivots can take a lot of forms: Getting more education, changing industries, moving into other specializations within an industry, or even complete changes (like you went to engineering school and now you want to open a pizza restaurant with really bad pizza pi puns, and in that case, you're welcome for the inspiration for your menu naming opportunities.). We can't guide you through what your pizza menu should look like, but we have made some changes ourselves, and we're happy to walk you through the steps we use to make changes in our lives.

First of all, what is the change you're looking to make? For the sake of this book, we will break some common pivot points into categories and hit each one in detail. In general, the career crossroads you may run into will be:

- Education
- Career changes (either changing industries or specialties in an industry)
- Promotions/responsibility changes
- Starting over completely
- Things you had no idea you'd ever need to consider

Let's start with education. You may have been like me and known a master's degree or beyond was in the works for you since high school, or you may have hit your college graduation with less than zero intention of ever hitting the books again, and that's okay too. You may have also gotten to the point you're at in your career with a combination of experience and education, even if not "traditional" degrees, say as a self-taught programmer or something. This case is less common in general, but especially so for roles that fall under more traditional engineering disciplines, like civil or mechanical engineering. In any case, you're now wondering if you need to take some steps to further your education to get where you want professionally, and like all good engineering questions, this one has an answer too. If you've never spent much time around engineers, then you may not know the first answer to virtually every question you give an engineer is some form of "it depends." In this case, when it comes to education, it depends.

The "it depends" for education is going to be based on two major questions you'll need to ask yourself: (1) Do I need this education? And (2) what is the expected pay-back for me? If you're considering doing a degree that costs you $100,000 but only helps you earn $5,000 more annually, it's on paper a bad decision. Like we said before, we want you to succeed, and financially is an important way to succeed, at least in part because it gives you the freedom to live your life how you want to, without always being indebted to the bank. We'll come back to this one.

How do you know if you need the education? Well, if you're reading this book early in your career, like before you've even graduated from college but you know you're going to want to do engineering, the answer is "most likely." We say that because there are always stories about someone who got into engineering just based on a combination of luck and experience and has no formal education, but we wouldn't want that to be the plan you're banking on. The more likely case is you will ask yourself if you will need additional education later in your career, possibly being just a year out of school or fifteen years out, whichever the case may be for you. You're probably going to be interested in more education so you can progress in your career, whether you want to qualify yourself to go deeper into engineering, or a pivot into something like management or possibly even something more different entirely. We have seen the gamut of education in our careers, from very successful engineers with only a bachelor's degree to much less successful employees with PhDs, and the inverse is also true – we've seen wildly successful PhDs too. All of which to say, the answer you get will depend on several factors. Fortunately, you can control a lot of them.

Wanting to learn more about a topic, refine your skills, or even pivot to something new entirely are all great reasons to pursue additional education, but in our opinion, they have to serve a purpose to you, either financially or career-wise. There is nothing wrong with education for the sake of education – in fact we will usually encourage it – but that doesn't necessarily mean you need to do formal education or take on additional financial burden to do so. Depending on what you are interested in, you may have a lot of non-traditional avenues to pursue this education, ranging from your employer to public resources. All this is great, but we have yet to answer the question we posed: How do you know if you need the education?

The answer to that question, in our minds, comes from simply asking yourself "how does this get me farther than I currently am?." If you have some work experience and are really interested in staying within the engineering field, then it probably makes great sense for you to pursue graduate degrees and professional certifications if you are interested in them. If you are looking to make yourself stand out from your peers, or open additional professional avenues sooner than you would otherwise, additional education can be a great way to do that also; a lot of positions consider a relevant graduate degree as two additional years of work experience, so if you have three years' experience plus a graduate degree, then presumably you are eligible for positions requiring five years' experience. Of course, this question doesn't only pertain to graduate degrees. I (John) had classes during my undergrad with a student who had more than 20 years of engineering experience, but finally wanted to earn his own engineering degree. Ostensibly, he had this same question himself, and realized this level of education would get him farther than he could get without it. The entire goal of additional education is that it needs to satisfy a purpose for you, or else it just becomes a hobby. Hobbies are great, but you need to make sure you have the resources for them. Also, if you feel like an impostor with one engineering degree, you'll feel that way with two or more graduate degrees, no matter how much you try to convince yourself that won't be the case. I (John) know this from experience, and we'll address some of that in the next chapter, and what you can do about it, because there is no getting around that feeling – you're going to have to go *through* it.

The primary issue with additional education is balancing the cost with the benefit. If you are in a position where you've left school but aren't landing jobs, our first suggestion would not be to just go take on more debt (presumably) and enroll in a graduate program in hopes that will make you more of a contender as a candidate. This happens sometimes, but again, is not something we'd necessarily recommend as your strategy. We have known people who did this, and it does work out for some of them, but in our minds, it's a big risk you may not need to take. Why is this a problem?

Because you have spent the cost, but you've potentially not realized any benefit. Another issue is sort of a false dichotomy – you believe additional education will help you advance in your career, but you picked some focal area that doesn't help you, or worse, qualifies you for something you don't even want to do. You'll want to make sure that any education you pursue actually helps you get where you want to go. For example, if you don't want to be handling the business side of the world, then earning an MBA was maybe not as helpful to you as pursuing a graduate engineering degree. That's not to say that it doesn't have the potential to still benefit you, but education is a tool for you to use to drive your career in the way you want. You have to use the right tool for the job, and that itself begins with some thinking.

The second major question we feel you have to answer involves the payback. It goes without saying that education has costs associated with it; you will spend both time and money on earning it. (A quick transition here: if you were in a position with school where it was truly a money-free equation, so you won't spend money earning it, and truly the time is the only thing holding you back, I (John) want to introduce you to one of my favorite sayings: "Time passes either way." Is time the only thing holding you back from something? You'd do a graduate degree but it'll take you two years? Time passes either way. You'll either be two years older with a graduate degree, or you'll just be two years older. I encourage you to apply this to a lot of things in your life, especially with ones that have little to no monetary cost associated with them: You want to learn a language, write a book, learn to code, take a pottery class, or learn some classic French cooking? Time passes either way.) To make a second quote of my good friend John Elam, you've got to "Do the thing."

Time aside, education has a monetary cost. For it to be a good investment in your resources, it has to pay you back. That is to say, it has to make you something in return. Graduate degrees are to me a neutral tool; they can either help you or hinder you depending on how you use them. I (John) pursued a graduate program about 18 months after graduating with my undergrad, and the program I chose cost me less than $32,000 from start to finish, but when I was done, I parlayed that into making an additional $40,000 annually, so in my case it was an absolute no-brainer. I didn't know that was the job I would pursue though, obviously, and I still chose to do a graduate degree. Why? Because the expected value of my degree was there. I originally wanted to find a degree that would pay me back over the same length of time it took to earn it; that is to say, if I spent $30,000 over 2 years to get a degree, I wanted to make at least $15,000 more a year in order to have it pay for itself. You may have a different scale or set of expectations for your additional education, but you're still going to need to know how you'll recuperate the cost of earning it.

The good news for a lot of engineers is that it is pretty common for even smaller companies to help you pay for continuing education, ranging from a few thousand dollars a year in smaller companies, to more than $25,000 a year in certain mega-sized corporations. This financial help, along with some planning on your part (read: Not going to the single most expensive institution you can find just for the sake of it) can really help you earn an additional degree with less or possibly even no cost to you. The catch to a lot of these programs is that if an employer pays for some portion of your school, you typically are contractually obligated to stick around for a period of time, or you may owe some or all of the money back. A pretty common agreement around engineering is that you agree to stay for a period of two years after the last payment for school is made, or you owe your money back. Some employers go less than the two-year mark, while others go beyond it. In our experience, two years is going to be a pretty typical timeline. Generally speaking, though, that clock starts from the last payment made, so keep that in mind when you're planning things out. You'll need to consider all the factors around where you live and work and if having to stay is worth it, but there are some paths and options for you to get graduate degrees at a lower cost.

We mentioned earlier that if you do pursue graduate degrees, we recommend doing them at the biggest name institution you can attach yourself to, and you may think this is counter-intuitive based on what we just said about not going to the most expensive schools around, and the answer is those are not necessarily related. I (John) earned my graduate degree from a very well-known and respected engineering school in Auburn University, and moreover the university had an excellent program for what I wanted. That being true, it was still only around $30,000 for me as I mentioned earlier, which means it was not one of those schools where you spend $100,000 to attend and come out with limited job prospects. In our opinion, that exact situation is fortunately limited for engineering students, as a lot of schools (including the cheap ones, the private ones, the public ones, the expensive ones, and everything in between) often produce engineers with great knowledge and opportunities upon graduation. The reason we say to go to a large school for your graduate degrees if you do one, is twofold: One, (1) the schools are well-known for a reason and they typically have larger endowments and research budgets, which drives reason number Two, (2) more instructors and opportunities for student aid in the form of grants, teaching opportunities, etc. We are certainly not saying that small schools have bad programs or instructors – not at all. They can have their own advantages too, but research for engineering programs drives a lot of things, even unintentionally. The main thing for you to focus on is that you pick a program that is a good fit for you, and that may take some digging. Finding a program that aligns with your goals and interests is well-worth the added effort you'll spend up front. Just because a program has the title of what you're going after, doesn't mean it is exactly what you're looking for. Programs vary widely from school to school, even within the same subject area. For example, not all schools that offer programs in aerospace engineering focus on hypersonic or vertical flight, so some programs may not offer you everything you want. The faculty and research from one institution to the next can vary widely too, so make sure to thoroughly vet out the programs you're interested in and pick the right one. Taking your time here is more important than just picking a school as quickly as you can.

Wanting to make some changes in your career is another reason you may find yourself at a crossroads, and we've been there ourselves. You may want to transition into a different role in engineering, pursue the management-versus-technical track (or vice-versa), or be thinking about leaving engineering altogether. As we said earlier, one of the best things about earning your engineering degree is that is a demonstration of problem-solving in practice. You could eventually decide you don't want to even be an engineer, and that's okay too. As we've said, we know quite a few people who left engineering for other pursuits, and have made quite fulfilling careers for themselves. It's your career, and you can drive it wherever you want to go, as long as you know how to steer it. Don't feel bad if you aren't on the exact path that you thought you would be on when you were younger, and don't feel bad if you're on a path that isn't even charted yet. All of these things are just crossroads you'll hit, and it's a part of life. We'll talk through some of the common paths you could end up on and give you something to consider.

Let's say that you took a job in test engineering because that was what you had available to you when you were job hunting. You've been here for some time, and while the job isn't bad, you're starting to think maybe you'd like to be in another area. After all, you've been learning and growing, and you know that any experience is good experience when you're starting out. Now at least you know that you don't love some aspect of the job you have, and that is something you can carry with you when you go look for your next role. The entirety of your career now likely won't be spent at a single employer like it may have been a generation or two ago, and that's okay. It's perfectly normal to make changes in your career, even if that means changing employers or industries. In fact, I (John) have changed industries several times, and have continued to refine what I am interested in and looking for every time. I am extremely grateful for the lessons learned in each of those industries, and can honestly say that I would not be the engineer I am today without those lessons learned at the right time for me in each of my prior professional experiences, regardless of industry. There is still quite a lot to be said for trying new things and taking some chances on yourself from time to time.

How do you know if you are ready to make some changes and pursue other things? Sometimes it's just obvious you need a change, and other times it's not. Maybe you are considering a complete career change, or maybe just taking on other responsibilities where you are now, and both are completely valid reasons to consider changing what you're doing. Career changes can come in all forms, and we understand that what motivates you to look elsewhere may not be what motivates another person. For that reason, we highly recommend that you have a good understanding of your reasons for wanting change. In some cases, you can get what you are looking for without making huge changes unnecessarily. For example, if you just don't want to be in test engineering anymore, you may not need to completely change companies and move across the country (or the world). You may be able to simply change roles in your current employer, or if that's not an option, other companies within your geographic location may have something for you. Change doesn't have to be drastic, but sometimes people need to make big changes, and we understand that as well. Sometimes you just need a new company, new industry, new city, or sometimes all of these at once. As we said, the first step for you is to understand why you're wanting to make that change, then you can take the subsequent action necessary to make it happen. If you're in the situation where you need a big change, you'll be interested in our "Transitions" chapter later on in the book.

Promotions can be another reason you'd like to make some career changes. Either naturally, in the course of your career, or because you took the initiative and wanted to go higher, you're going to get promoted eventually. This in itself can be a driving force, and you can leverage it into big changes, like moving, funding additional education, or the like. We've already talked through promotions and how to get them, so we just mention here that they can be a significant driving force in your pursuit of change.

So far, all the change scenarios we've listed have been more or less something you chose to do, at least in some capacity. Okay, but what about when something completely unexpected and utterly out of your control happens and you become forced to do a complete 180° turn? Even if that means starting over completely? This scenario is very rarely a fun one. Honestly, this scenario is probably never fun, nor is it generally predictable. Let's say you lost your job from any list of reasons: The company folds and goes under from decisions made by bad leaders at the highest level, a worldwide pandemic makes entire industries go away, or you end up with some kind of illness that forces you to stop working as you intended. All of these things can, and sometimes do happen, and the first thing we want to do here is say if any of those categories are one you belong to, we're sorry to hear that. One of the elements of the human condition is that we are often left wondering why things happen, and we may never have answers to that question. If you found yourself in any kind of situation like that, it's okay to take some time to grieve what happened. It's okay to not be okay in those moments, and it's okay to not know how to move forward. Begin with taking some time to process what you need to, and then develop a plan to move forward.

Once you start moving forward, you'll need to identify your needs versus your wants. If you're in any sort of situation where you don't have the money set aside to weather this storm, then your response will necessarily be different from someone who does. You may even be in a situation where you need to start making money sooner than later, so you may not get the luxury of being ultra picky in your search. That's not ideal, but it's not permanent either, so you can always change things around when the opportunity presents itself. The next thing to do should you find yourself in this situation is start looking at what is transferrable in your career, and compare that against the scenario or environment that landed you in the position you're in.

That was a lot of words, so what do we mean here? If your job was impacted by an automaker going under, but the industry as a whole is still doing okay, then that's a totally different situation than your job went away because you worked in cinema, and a pandemic made the entire cinema industry go away (looking at you 2020). Both end in your job being impacted, but the way you get out of each scenario may be different. This assumes, of course, that you wanted to stay in the same industry to start with, and weren't already considering a switch. If you were considering making a change, these kinds of life events can be great catalysts to force you to make changes you otherwise wouldn't have.

Back to transferring your skills and moving forward – To the extent you can, you're going to need to determine what you take with you to a new job. These crossroads, like all others in life, come with some obvious questions, but also with an iceberg of unknowns that maybe you haven't thought about or even realized were a potential deciding factor in your choices. For example, when I (Sean) was about four years into my career, I realized the position I held was not going to provide me the growth necessary to facilitate my life outside of work. I had just bought a house and was planning on proposing to the woman who is now my wife, and was stuck making the same wage in the same position I had been in since my second year of my career.

I knew something had to change, but then a position opened up across the country in a place I had never been, with a company I knew of as one of the greats in my industry. The job was a promotion, the pay raise was astronomical, the experience to gain from these other engineers in a department ten times the size of my previous one was immense, it all made too much sense not to take the dive and move. That is exactly what happened, and my wife and I made the choice to pack ourselves up and drive 2000 miles across the entire United States to our new lives and my new job that was just going to be amazing.

Except, it wasn't. Don't get me wrong, the job was pretty much exactly what I expected and I learned quite a bit of how a larger organization optimizes their engineering as opposed to the more "Mom and Pop" way of doing it I was used to. But some problems still remained. First, the move was not the perfect fit my wife and I expected it to be. It took a few months before everything made sense but we quickly realized a few things. One, the location was not for us. The United States is vastly different when driving even 200 miles from one city to another, but 2000-plus miles provides a whole new sort of people and locale that has to be experienced to appreciate. We liked traveling around and visiting everything around our new homebase, but long-term living was not meshing with the lifestyle we liked to live.

Second, a promotion does not necessarily equal an increase in responsibility. As one of two engineers at my first company, I was the end-all-be-all of multiple programs. If someone had a question, I had the answer. If someone needed something done or a customer requested a change, I was the guy to make the changes and take charge. However, switching company sizes resulted in something strange. I was now in the back seat, designing one small part of a much larger piece and the layers of bureaucracy on top of it made everything feel like a snail's pace. This is normal, as most large companies cannot make large sweeping movements in any short amount of time, but for me, this was new and sort of disheartening. I had thought I made a move up, not down and to the side. The third thing I did not consider is the interpersonal implications of moving so far. What do I mean? Well, my wife, bless her heart, dropped everything she was doing and moved with me which leads to its own stress in her life and therefore our life. Additionally, I had left a very strong family and friend dynamic and started to work to rebuild something similar in a new place which as stated previously, we did not mesh with. This all amounted to a work life that was not as expected and a personal life that had to be worked on at a level we had not anticipated. All of this resulted in us leaving to go back from where we came from after only a year in this new role.

The examples above were my unexpected deciding factors in the job shift, and while some of them may ring true for your situation, in the end, this is your life and your career and what drives me may not drive you. Crossroads are inevitable and change is good for the soul, but be sure to think through your choices fully or you may end up spending 5-figures moving 4000 miles in a year like this author did. As a note to consider in wrapping up this section, things probably won't be exactly as you anticipated them being, either in taking new jobs, moving across the country, or in anything else for that matter, but there is still a significant silver lining to be found. While Sean and I had different (but very similar) experiences at the job he moved across the country to take, it is that job that brought us together as authors and lifelong friends.

That is to say, there is both good and bad in places you don't expect it, so take what good you can and understand the rest may not be for you, and that's okay.

CHAPTER SUMMARY

As you develop and grow, you're going to come to a lot of crossroads in your career. That's completely normal, and isn't a sign that things aren't going well. Often, it's the opposite of that. As you grow, your goals and even your circumstances will change. As you navigate those changes, it's important that you keep a few things front and center in your decision making:

- Things (your career, education, and life as a whole) probably won't go *exactly* as you planned earlier
- Know your goals and what drives you, as that will help you plan how to get there
- You can't control everything, and sometimes you'll need to adjust to the unexpected
- Sometimes the biggest source of growth happens when you least expect it
- Time passes either way, so unless there's a good reason not to, you may as well just do the thing

If you can at least know your goals, you can keep adjusting to drive toward them, and that's really the whole point. Your career most likely won't follow a perfectly straight path from beginning to end (ours sure didn't), and that's completely fine. At the risk of sounding too cliché, life is absolutely about the journey and not only the destination, so we want you to be able to enjoy the ride as much as you can, and really drive your career where you want it to go. We want you to be able to live the life you want, and we want you to be able to help others do the same, so our hope is that we've helped equip you with the things you need to navigate the early stages of your career.

8 Fighting the Impostor

At some point in your career, whether it happens at work, in school, or is even a recurring theme you keep getting stuck on, you're going to feel the effect of what is commonly called "Impostor Syndrome," and you're going to feel that, quite frankly, you just aren't good enough to be doing what you're doing. This is something else that is not only extremely normal, but that hits us all eventually, so you're not alone in this feeling. It's a part of learning, and that makes it a part of life – you'll see someone else and say to yourself "they are so much better than me" at a few things, or maybe a lot of things. Recognizing this and knowing how to deal with it is also a part of life, and is in our opinion, an extremely underrated skill.

I (John) still remember the first time I felt inferior academically: It was my second to last semester in my junior college, and I had made my first D grade in my life: Calculus II. I am horrible at Calculus because I was never good at algebra (but at least I was good in trigonometry, and this single handedly saved me in college). I came from straight A's my whole life to the occasional B or C in junior college, but that was more a product of me not putting in any effort and needing to learn better approaches to dealing with things I didn't want to learn than it was me not getting it. Calculus was different. I did okay in Calculus I, but when I made it to Calculus II, I was lost. To this point in school, I was great at anything trigonometry-related, and less so with algebra. Ironically, it was the hyperbolic trigonometry functions that completely derailed me in Calc II, and I knew I wasn't going to recover that grade. It was the first class I had to retake, but it unfortunately wouldn't be my last. I did some arranging and retook Calculus II *after* Calculus III and differential equations (I made a D in the fall semester, and took Calculus III and differential equations together in the spring semester – which along with Physics II and a lab, made for a combination of classes I would never recommend you do – so I retook Calculus II over the summer). I passed and life was good again, until I left my junior college and made it to my university, where I encountered Statics, Mechanics of Materials, Thermodynamics, and still a couple other classes that I ended up having to retake. In fact, somewhere in there, I had an instructor who told me during a visit to his office hours that I was too stupid to graduate, and that has been a conversation from school that I have never come close to forgetting. I made a hole for myself over time, and I had a mess on my hands. Eventually, I was able to get out of it, and I did graduate, but not without a lot of damage being done to the way I saw myself, and how the school saw me for that time. As you read at the beginning of the book, I got kicked out of my department and nearly the school as a whole. Ultimately, I recovered from that point and graduated, but the injury to how I saw myself was already done.

DOI: 10.1201/9781003510901-8

Fast forward from there a couple years, and I was leaving my first job. I had gotten an opportunity to go join a hugely known Bay Area tech company, and I was ecstatic. After graduation, I had gotten a job at a fairly small family-owned business that had made a name for itself developing the standards that grew their industry from a niche market to something that shows up in everything, but the company itself was not really well known, and there was not much space for me to grow there. This company had acquired another company that was close, and somehow, for reasons I will never fully understand, I got an interview there. Then another. Then they flew me out to their corporate office in California, and what's more, they made me an offer on the spot. They liked me so much, that they rewrote the job description to make it fit me better. In fact, the original job I applied for was a senior-level position that wanted at least ten years of experience, and I had about two. I honestly don't even remember seeing that it was a senior-level role, as I was applying to so many things, and had I seen that, I never would have applied. Trust me, when I say that unicorn jobs exist, but they aren't the norm at all. I had in the span of a few weeks gone from a small-time shop to one of the most recognizable companies on the planet, at least partly, according to one of the managers who fought to get me there, because of "the fire in [my] belly" he said. I was eager and excited, and I couldn't wait to get going and make my mark. I was ecstatic and soaring on my own success. I had gone from almost getting kicked out of college to landing this job. I was going to be a success story, and wouldn't even think about failing a couple of classes in college ever again. What's more, I'd never have to actually deal with it either.

It's that simple, right? Sadly, no. I had some more heartbreak coming my way. It wasn't long though, before I was struggling there in this role. Like REALLY struggling, to the tune of those managers who brought me on saying things like "we don't want to have to replace you" to me. I was crushed – no, I was more than crushed, I was devastated. What happened? How had I gone from the straight-A high-schooler to a struggling engineering student, and worse, a struggling employee? Was I an Impostor? Was I a failure? Do I even belong here? I didn't know how to help myself get better, but I did know something had to change, and change fast.

By this point, and a lot of self-reflection later, I had come to the realization that I was flat-out doing work that wasn't good enough, and I had been doing it since college without even realizing it. In school, either from laziness, disinterest, or some other cause, I was coasting by on the minimum effort. My first job out of college was so easy I didn't have to change my ways, and moreover, nobody there cared enough about me to care. In that job, everyone was committed to the status quo, and my management didn't care enough about me to help me grow, or show me where I needed to grow. When you're starting out, you don't have the ability to proofread your own work, and you don't know what you're looking for when you have an introspective. You naturally tend to be very myopic and ask yourself "what would make today be done faster?" This is one reason why it is so important to be able to recognize the traits you need a manager to have because when you are starting out, you need someone other than yourself to point our areas to improve. Much like how it is possible to teach yourself guitar on your own, but you'll probably learn faster with a good instructor, the same is true for your professional development. You are going to need people in your corner who will help you, especially when starting out. You need people to coach you into getting better, and that's why we push mentoring so much because you don't always get the coaching you need from your manager or those in your direct leadership chain. If we can help you avoid (or at least leave you better equipped to recognize and manage) the behaviors that led to our own struggles, we'll consider this book a success.

Now six-months into my tenure at this tech company, I had burned through all the new-guy good will I had, and things were going downhill. The same questions I had been dodging since Calculus II reappeared: Am I a failure? Do I belong here? Why isn't this going how it is supposed to? Turns out, I wasn't a failure. I was just misguided with my efforts. I honestly didn't know how to approach the things in my career, or going back to my school days, I didn't know how to approach things that just didn't click at first. That is something I had to learn, and unfortunately, I made that lesson about as hard on myself as I possibly could. I needed people to sit me down and say "John, this isn't working." My time at this employer, and more importantly the two managers I had there, are unequivocally the reason I am the engineer and person I am today. Trevor Davies and Aria Liaghat, thank you both for teaching me in spite of me. You both pushed me to be the best version of myself I could be, and I owe you for saving my career. I had a lesson to learn, and it was not a fun lesson – at all. This tech company gave me the opportunity to forge myself into what I wanted to be, but forging isn't without trials. For steel to be forged, it must be heated and shaped – usually with a giant hammer! I had a lesson to learn, and every single person of the 40-some-person team I was a part of helped me get there. Over time, I began to see myself differently, not just professionally, but personally as well. My wife Monica still tells me how different I am as a person from when we started dating and how much I have grown. The takeaway from this chapter is that lessons are always worth learning, even if they are difficult.

Fortunately, I learned this lesson early enough to turn the ship around, and things worked out quite well from there. I had a luxury on my side that not everyone has: My employer was willing to teach me. They wanted me to be at a place, and they decided to help me get there. They had patience and resources to help me right the ship. I struggled for about 6 months, then worked hard to correct what needed fixing for about 6 months. After that, I spent the next year showing everyone that I had not only learned from my mistakes, but that I had grown, that I was capable, and that I was the right person for the job. Lessons can take a long time to get through, even with the correct amount of effort. Slowly but surely, my job there transitioned from me feeling like I had no idea what I was doing, to then feeling like I was seeing the light at the end of the tunnel, to ultimately being "the guy" that people knew they could count on. I left that company and position with a lot of good friends, and a lot of respect for those who helped me turn the ship around, even if sometimes they seemed hard on me. It was absolutely for my benefit, and over time, I stopped describing myself to myself as "in over my head," or "not remotely qualified to do this job," and gradually shifted to a view more along the lines of "I've got this," and "I've fixed this before." For reasons even beyond this anecdote, my time and my team at this tech company will always be among my most cherished working experiences. I cannot overstate how grateful I am to have been on the receiving end of a company and people who cared, and it is not an exaggeration that those two years saved my career.

The concept of Impostor syndrome may possibly be a little fuzzy to some because they can't quite define it, or how they fit into it. You're never really confident in your abilities and aren't sure if you are ready for the responsibilities you've taken on, so you feel like an Impostor. I think Impostor syndrome is very much like a coin, and that is one side to it. The other side of this Janus is something entirely different, something like a "pseudo-Impostor" where you feel like you're "there" but you've never really arrived. You're not putting your best foot forward and you've done it for so long that you don't even realize it anymore. I was the latter – I had become something I didn't even recognize anymore, so by definition, I was an Impostor.

I say these two stories about my own failures in college and work to illustrate a point that may be tough to hear: Sometimes you feel like you don't belong because you *don't*. At least, not the way you are acting. When you feel like you are in over your head, or that you are an Impostor who doesn't belong, I want you to reflect and ask yourself "Am I putting in my best effort?" Until you put in major league effort, you don't deserve to play in the major leagues. Maybe you stopped putting in effort because like me you didn't have to for a while, maybe you were too proud to admit you were struggling, or maybe you felt like you didn't need to work harder, but somehow you found yourself in a place you didn't like, and it's time to change that. We're going to go into more details and specifics on how you can help yourself out of those situations in our next chapter, but first, I want to make sure we can all recognize this for what it is: not putting in effort. You will feel like you don't belong where you are when you're not putting in the correct amount of effort, and that will catch up to you eventually. Even if it takes a long time to catch you, there is simply no outrunning that fact. If you're lucky, you learn this lesson when you're in a place where the company or school or organization is ready and willing to help you. If you're not, then this lesson is going to be even less fun for you to learn. We want you to succeed, and we stress this over and over. Talk to your peers, find mentors, take the instruction and correction from your teachers and managers in the spirit it is intended, and above all, don't just careen down the path of not improving or just maintaining the status quo.

Nobody knows you better than you know yourself, so take time to reflect be honest with yourself: Do you feel like you don't deserve to be here because you know you don't? That can be fixed, your situation can be improved, and you can make things work out, but you have to recognize you're in a hole, and stop digging before you can get out. It doesn't make you a failure or a waste, but rather someone who needs to admit where they are coming up short and start making efforts to correct issues. Your entire career and well-being will thank you for making these changes now, and ideally, before it's too late. You don't want to find out that you can avoid your situation by putting in the correct effort or doing things differently when your school or employer decides it's time to part ways.

Obviously, this doesn't mean that all cases of Impostor syndrome are caused by you not putting in effort, nor does that mean you'll stay in that category forever. There are times when you will feel like you don't belong even when you are putting forth maximum effort, and you are meeting the requirements of the job you have. This is the very real and traditionally recognized feeling of Impostor Syndrome, and it's wholly different than what I described as my story for the last couple years of college and shortly after. I even experienced this traditional Impostor syndrome feeling myself *after* I left the tech company and went to my next job. I felt like things were running *too smoothly* after the previous couple years, and that was just due to the fact that I had learned and grown so much over my time there. It is entirely normal for you to go through phases of one or both cases, and sometimes you'll go through them more than once or in different orders. The goal is to recognize what is in front of you, and continue to keep growing as a person, a student, and a professional so that you can continue to adapt to these challenges and overcome them. If you're confident that you are already working at a level of competency that you can claim is your best effort, but you don't feel like you belong, then you fall into that classic case of Impostor syndrome.

The traditional case could be caused by you listening to nagging doubts, something someone told you once, or even that you're just less experienced than your peers (or even from pejorative views/comments/opinions of others around you, and we want to state that engineering is for everyone. If you've earned your way into the ranks to be qualified to be one of us, then you *are* one of us, regardless of particular schools, degrees, or anything else). Whatever the case or cause, you find yourself in a predicament that's starting to creep into your thoughts and it's impacting how you see yourself, and maybe even the work you do, so it's time to address it. Growing as a person and running into, then later through, Impostor Syndrome can be a delicate balancing act trying to balance pressuring yourself to get better and hit your goals while also acknowledging that you're only a human and will never be perfect. This balance can only be found after you've had an honest, very frank look at yourself and assessed where you are effort-wise.

We've already alluded to their names, but we'll go on and name the two types of Impostor syndrome we've outlined here for the purposes of the book and conversation. In the first case, where you're not putting in the effort you need to and that impacts how you see yourself, we'll call "Effort Impostor," and in the second, the more traditional feeling of not belonging, we'll call "Classic Impostor." Once you have had a serious moment (or a few) of self-reflection, you can decide if you are dealing with a case of "effort" or "classic," you can determine your next steps.

A point here; self-reflection is a crucially important component of your professional development (and even personal), but that doesn't mean it has to be "lonely-reflection," meaning you shouldn't only consider your own opinion. You should be actively seeking feedback from others regarding your performance, and this is something I have seen organizations do both extremely well, and extremely poorly, so you need to know how to solicit feedback for yourself. A lot of people go into the workforce believing that their boss or manager will provide them with all the feedback they will need, and that's maybe a little too idyllic for the real world in practice. Your direct manager is there to help you and provide guidance and feedback, yes, but there is a very real possibility he/she will be less effective at this than you may need, for a combination of reasons, ranging from they are too busy to the unfortunate fact that they may just not be very good at giving feedback. I highly recommend you start working on soliciting feedback and developing it into a regular part of your career, especially in the earlier days before you've gained a lot of experience. I (John) have used the following method since my days at the tech company, where they helped me turn the ship around, what I call the **"Plus/Minus/Delta"** method: I ask my peers (a broad term, but one I use to mean both those doing the same job function as me and those doing different job functions but with whom I work with closely), my seniors, and others who touch my projects (I have spent my career as a project engineer/manager, so I had the benefit of knowing a lot of people) to provide me feedback. This feedback is simple, and follows a three-part flow of

"Plus"— things I do well;
"Minus"— things I don't do well; and
"Delta"—things I could do differently to make their jobs easier or better.

I solicit feedback following this format because it's simple, and because it allows people to feel some confidence in telling you what you can improve upon by letting them tell you what you do well. It's also important to consider here that good, constructive feedback is often not all that easy to get, partly because it is often hard for people to give constructive criticism. They can either feel poorly positioned to review your work, or they just do not like criticism themselves, and that can make getting good feedback hard. What makes good, constructive feedback? It falls into that broad category of "you'll definitely know it when you see it," but in general, it has to be actionable –meaning it has to be something you're capable of changing – and it has to be given with the intention of helping you get better. It does *not* have to be easy to hear necessarily, because sometimes the truth isn't fun, but it has to be from a place of building you up, thus being "constructive." For example, someone saying "I don't like your voice" is not constructive criticism to your public speaking, whereas someone saying "you pause and say 'umm' too much" is something you can work on.

This concept is important to understand because you're always going to be pursuing improvement, but you're also one day going to be in a place to help others improve, and you want to do that in a way that doesn't belittle them. Say what needs to be said, certainly. But say it in the *way* it needs to be heard, and in case you are wondering, that won't be the same way for all people. It's going to take some skill and some learning to know how to deal with everyone you'll encounter, as well as learning what feedback methods, styles, and tones, work for those around you. Back to the method: Asking for the things you do well is a great start because it lets you know what you can continue doing, as well as giving the person from whom you are soliciting the feedback a way to not seem overtly negative, and that's why you start with the "plus."

The next component, the "minus," is the part that may not be fun to hear, but will be things you need to work on. That's alright. We all have things we need to work on, or that we have needed to work on in the past, and that's okay. You may be an engineer, but you're still human. Since we already outlined constructive criticism before, here we'll just say that not everyone will give constructive criticism, and they'll only focus on the criticism part. Consider the context, the person giving the criticism, and what is said as a whole together, and generally try to only get criticism from those who can speak to your work. I once had criticism given to me from another employee by a direct manager, with permission of the employee. I knew it wasn't going to be fun to hear when my manager apologized for the harshness of the review before he even began reading it, but the intent was to get me performing not only to the level I needed to be, but to the level they knew I could. His tone was kind of harsh, but his words were exactly what I needed to hear. "Minus" feedback will be the kind that we dread the most, but it will also be what develops you the most.

Finally, we have "delta." Delta is going to be helpful to you. It opens up a door for people to say the minor things they haven't wanted to bring up because it's not *that* big of a deal; things that help them with their jobs. As you are working, it's helpful to remember that everyone has a job supporting other people – as you get your work from someone else, so too will your work go to someone else, meaning someone down the line picks up where you leave off – and it's a nice touch to hand off work in a way that allows others to just do their jobs, not make corrections to yours. As an example, "delta"-type feedback could be that you often generate a report on a Thursday afternoon, but the person who needs your report has to present on Friday mornings so it's often a scramble to get your work in. Your work isn't wrong per se, but it would help this person be more effective at their jobs if you can deliver this sooner, say on Thursday morning instead of Thursday afternoon. You'll find little changes in your habits can really impact the quality of your work, and in turn, the ease and quality of your coworker's outputs as well.

We would recommend that when you are starting out, you solicit feedback multiple times a year, probably quarterly at a minimum. Your employer may have really good systems in place to help you do this, and they may not, but you need to solicit feedback early and often. You may not need to do it quarterly forever, and once you get a bit more experience, you'll find that every six months or so works well. The other underappreciated benefit in soliciting feedback regularly is that you'll have a clear line of sight into what you need to do to be ready for more responsibilities, promotions, and general growth as you progress in your career, something that you'll certainly want to have if not now, at some point in the future. Good feedback can serve as a road map to taking your career to the next level. I have vaguely outlined what good feedback looks like, so I want to dive a little deeper here and help you get an idea of how to provide feedback. Soliciting this feedback helps you in other ways too, especially if your organization isn't the best at giving it to you organically to start with. Ideally, the first time you hear about the potential negative aspects of your performance won't be during your annual review period, but sometimes that happens. If you find yourself in a situation where you didn't find out about things you could improve, then it's generally a reflection of management that has room to improve itself. Soliciting this feedback on your own helps you fill in the gaps potentially left by your own management, and it helps you be better prepared to address things while you have time.

As I mentioned a little earlier, you're always going to be around feedback as a professional, either as a receiver or as a giver of this feedback. Something I think a lot of people struggle with is how to give feedback or review people, so they come in and say "I think this person is okay," then proceed to hem and haw but not really say anything. How do you give good feedback? Keep in mind that it has to be actionable, it has to be constructive, and it has to be relevant to the person and position. If Bob is a software engineer and he makes terrible chili, that's not relevant. If Bob makes a lot of mistakes from lack of attention and his code isn't following standards, that is a problem that can be worked on. Good feedback will always be based in tangibility, meaning it has to be something that the person can change, and it has to be something that is based on reason. So many professionals don't know how to help people improve, so what ends up happening is that nothing changes and everyone flounders. If you are providing feedback, one of the best ways is to begin by asking yourself what would have helped you develop if you were in this position. For example, you've been asked to help with mentoring a junior employee in a similar position to yourself, so you begin by reflecting on things you wished you knew earlier. There is often no better way to learn the material more deeply than by teaching it, so if you're ever in a position to mentor more junior employees than yourself, you can really find opportunities to learn the subject matter as deeply as you want, and that can further benefit your career development.

> Okay, great. I've taken a good, honest look at myself and I know I am putting in the effort I need to be, and I the feedback I've been getting says I'm on the right track, so why do I still feel like I'm not doing well enough?

I'll tell you why – you're experiencing the classic Impostor. This can, as we said earlier, hit you at any point in your career, whether academic or professional, and may even hit you multiple times as you grow and learn new things. Experiencing Impostor syndrome is normal, unfortunately, and it doesn't mean you are bad at things or unqualified to be where you are. What it does mean is that you need to evaluate yourself and see why you're feeling this way and take the appropriate steps to identify the cause. Once you're sure you aren't in the trap of the Effort Impostor, look at why you're experiencing the Classic Impostor syndrome. Have you taken on new responsibilities? Have you gotten a promotion or joined a new organization? It can be intimidating to compare yourself to others, especially ones who have slightly more experience or were perhaps in some way better suited for their roles than you are. The classic Impostor is a completely normal and common thing too, and if you know how to make the most of it, it has the potential to be useful on its own.

What causes the classic Impostor syndrome for engineers, especially? A lot of things do, but to us, really two in particular stand out. Humans are wired to be social creatures, and that leads among other things to comparison to others, so there's one potential cause. We compare with things and people we should never compare ourselves to, but we still try to reduce our environment to square pegs and round holes. A physiological phenomenon called the Dunning–Kruger effect is another cause of Impostor syndrome, and we think it is quite applicable for engineers in particular. In Dunning–Kruger, to over-simplify it to the point of paraphrasing with broad strokes, less intelligent people are more confident in their abilities because they don't know what they don't know, while very intelligent people know they don't know everything, and thus doubt themselves. In fact, they can even ultimately tend to doubt things they *know* they know. Taken a little bit farther, the German–American novelist and poet Charles Bukowski said "The problem with the world is that intelligent people are full of doubts, while the stupid ones are full of confidence." There is a lot more to it than just that reduction, but it'll serve our purposes here. Engineers, on the whole, are pretty intelligent people. We have to be in order to make it through engineering school, and all the classes associated. That simple fact, by human nature, means you are more likely to encounter some form of Impostor syndrome than you would be otherwise. We authors as engineers ourselves can guarantee you that you are not alone in feeling this way, and we, along with tons of our peers have felt this at times all throughout our own careers. The truth is that everyone who is honest in their introspective reflection has felt this at some point or another, so more than likely you're going to need a strategy to *deal with it* instead of trying to just *avoid* it. That at some point the Impostor will catch you is virtually unavoidable leads to the real question: What do you do about that? Is it always a bad thing?

The answer here may surprise you, and it's not necessarily a bad thing (mostly). The Impostor shouldn't be seen as an exclusively bad thing because you now know that you're pretty smart, you're capable, and since you've earned the right to be where you are, you belong there. If you know that feeling that way is normal and to be expected from time to time, then you can respond accordingly. The only time feeling some Impostor syndrome is truly bad for you is any time that you can't shake the feeling, and it starts wearing down your emotional well-being, or you start seeing yourself as a failure and only a failure. If you start to feel that way, we encourage you to again ask yourself "Why do I feel this way?" Do you feel this way because of comparison to someone else, or do you like me, often hold yourself to some kind of standard that is in no way attainable for *any* person to meet? Take a step back, talk to your mentors, and address it head on. You have to move past the "I'm a failure" stage to get to the part of this you can use to your advantage: The recognition of growth.

Often, you encounter the classic Impostor syndrome because growth is by nature, not usually a comfortable experience. Consider it – whether academically or professionally – to be an experience not entirely dissimilar to going through your middle or junior high-school years. You're growing, so you're not where you started, but you're not where you're going yet, so things can just be awkward – but, and we stress this, you are growing. That in itself is worth being proud of because of the progress you've made. If you're in college and feeling like an Impostor in the harder classes in your major, be proud of the fact that you've gotten there. If you're feeling like an Impostor in your first job, be proud of the fact that you've made it through engineering school and gone through the process of landing a job. If you're in a later position in your career and feeling like an Impostor, then you can be proud of the challenges you've overcome to be in that position as well. We say to be proud of your past success not to dwell on where you came from, but to put in perspective that you have continually found ways to overcome in the past, and you can do it again. The positive perspective that can come from feeling like an Impostor from time to time comes from recognizing that growth is a dynamic event, and it doesn't stop just because you have now graduated, or gotten a job, or been promoted, or whatever event you want to use. It can be uncomfortable because it is new and unfamiliar things often are uncomfortable, but in the case of you continuing to develop, this discomfort is the result of maturing and growing beyond where you started. Try to frame these experiences in sort of a crawl-walk-run mindset: You didn't know how to do the simple thing, then you did, and next up was the bigger, more complicated step, and so on.

Obviously, this positive mindset assumes that you really are putting forth the required effort and not just allowing yourself to coast by on hope, as was the case in the Effort Impostor. If you are certain that you are doing what you can, and you're certain that what you're experiencing is the classic Impostor, those feelings of discomfort are a surefire indicator that you are growing. We stress the recognition of growth because human nature is such that we tend to focus on how far we have left, but rarely on how far we've come. "But getting where I am going is my destination, not how far I've made it since I started!," you say. We hear you, and we want you to get where you're going too. By recognizing the growth you've already had, you can recognize what you've done in the past when faced with uncomfortable things, and formulate a plan to move forward again.

You've probably noticed by now that we are big proponents of self-reflection, and that's because life is a journey with many, many stops along the way. Self-reflection helps you to frame things in a way that focuses on what you can change, as well as see things for what they are. If you are feeling like you're an Impostor because you're not putting in the required effort, or you're slacking somewhere else, self-reflection (and a healthy dose of soliciting feedback) will help you discover that. If you're feeling like an Impostor because you're growing, and you are not the same person/student/employee that you started as, self-reflection will help you discover that too. The way you fight the Impostor boils down to knowledge: Knowledge of self, knowledge of growth, and knowledge of areas you can improve.

CHAPTER SUMMARY

If you haven't noticed by now, self-reflection is a huge part of growth, as is feeling like an Impostor from time to time. The two main types of Impostor syndrome are what I call the "Effort" and the "Classic" Impostor. In the former, you feel like you don't belong because you're lack of effort has finally caught up with you, and is going to take some serious self-reflection and changes on your part (but that doesn't mean it has to be changes you make alone!). In the latter, the Classic Impostor is a case of you growing and no longer feeling like you belong at the table, that you "outkicked your coverage" so to speak. This is entirely normal, and unlike the Effort Impostor, doesn't mean you need to make some changes. The only way to know which category you're in is for you to do some self-reflection, and solicit as much feedback as you can, which is a skill that pays dividends throughout your career.

Main things to takeaway here:

- Lessons are always worth learning, even if they aren't fun to learn
- Sometimes you feel like you don't belong because you know you're not working hard enough, and that's what we call the "Effort Impostor"
- Sometimes you feel like you don't belong because it's all new to you, and that is a sure sign that you're continuing to grow, and that's what we call the "Classic Impostor"
- You can feel either at various times in your career, but you won't ever be in both categories at once
- Comparison is the thief of joy, so make sure you are doing true self-reflection, and not comparing yourself to your peers
- Soliciting feedback is a great (and often underutilized) skill for you to develop in your career
  - One of the best ways you can drive this conversation is by using the "Plus/Minus/Delta" method, where you ask what you are doing well, what you could be doing better, and what you could be doing differently

9 Career Transitions
Changing Jobs, Industries, or Positions

If anything is certain, it is that in life, nothing is certain, and that applies to our careers as much as anything else. The tradition or even expectation of those who came 30 years ago or more before us, the ones who landed an entry-level position and leveraged that into a career climbing the corporate ladder and ultimately retiring with the same company is a distant image in the rear-view mirror. These days it is much more commonplace to become a nomad of sorts, transitioning from company to company throughout your career. The advantage of this being, of course, a massive increase in opportunities to you, since you no longer have to wait for someone to retire or for your company to decide you are ready to take a promotion and move to the next step in your career path. Additionally, it is possible to take what experience you have cultivated and apply it to perhaps a different role or industry and find a better fit for your skills and interests than what your company has available at the time. Being a career nomad has some downsides, among them the uncertainty of it all, starting again as the new person, having to decide when is the right time to move on, and discovering which of your skills best translate to the next stage of your career. As with many things, the good and the bad are present together, and with some knowledge and effort on your part, you can work to maximize the good parts of the career environment we find ourselves today.

Whether you plan to make a major career change or a minor move, you'll likely find yourself in this situation at least once in your career, and probably several times if we are fully honest. The best way to approach this situation is to identify the transition being considered and find the best time to make the move, as well as any potential challenges that will accompany the change. For the sake of our discussion, we'll define major career moves as those when you are looking to move companies within your industry and retain the same job title/responsibilities, move on from your existing job title into a different aspect of engineering (think a manufacturing engineer transitioning to a program manager or vice versa) either internally or externally, changing industries entirely, or moving on to a major promotion in your career path. A minor transition then, would be something like moving projects or departments or a standard promotion from Engineer I to Engineer II at the same employer. While exciting, these should just be a regular part of your work life and do not necessarily require as much forward thinking when accepting the change of responsibilities.

So how do we know when it is time to initiate a career transition? While there is no clear line to be passed or list to be checked off that causes you to shout "Okay, now I should move on!," there are definitely signs to indicate it just might be that

DOI: 10.1201/9781003510901-9

time has come to move on from your current career situation. Starting with what may be the most common career move – moving companies within your industry – we will take a look at what should be considered when contemplating this transition, as well as what signs you should be trying to find.

While in previous chapters we have identified pathways to maximize your growth and position yourself as best as possible within your career, your employer still plays a significant part in how your career can blossom – IF you let them wield that power on their own. Changing employers to another within your industry can be a great way to improve your income, grow your knowledge in your craft, and remove yourself from some negative aspects of your current work situation, and doing so grants you further control of where you will end up in your career. Speaking from personal experience, after graduating engineering school I (Sean) accepted a position at a small company with an engineering department of just three engineers. Obtaining this job right out of college granted me the opportunity to cut my teeth through every phase of the R&D process with heavy involvement that schoolmates of mine never dreamed of while at larger firms, where fresh engineers were hand-held through the more basic parts of the job. It was challenging to a degree no other position has touched for me, yet I was able to expand my repertoire and pack my resume with a level of experience usually reserved for engineers 5–10 years my senior. The downside to being a fish in such a small pond is there's a minimal pool of peers to extract knowledge from and ultimately no clear path for your growth. It is hard to be promoted when the only position above yours in the department is your boss's position.

Additionally, the department's smaller budget meant the pressure was on for programs to be on time and done right the first time to mitigate cost overruns. After a few years of this, an opportunity arose at a well-known company in my industry where just the section of the engineering department I would join was larger than the entire company I had been working at. By taking this job I was able to boost my income by 30% overnight with improved benefits, and I had the opportunity to now learn from other engineers who had been on the cutting edge for decades. In the end, I discovered the industry wasn't for me and transitioned to another and have been with the next company ever since. However, had I not made that jump I would have always wondered if I could keep up with the engineers who had decades of knowledge and millions of dollars of budget behind them.

If you ever wonder about this, we recommend you try it and see where you land when the opportunity comes. There is absolutely no shame in wanting to try something new and push your boundaries. If you try it and you don't like it, at least you know and don't have to wonder. We generally recommend this approach to life as well. If you want to try a different state (or city) for a while, want to give an industry a shot, or just need some change, we recommend you give it a try when and while you're able. If you happened to grow up, attend college, and now work within 50 or 100 miles of the same location, we encourage you to get out there and see what the world has – there's a lot more to the world than the same people you've known since you were in middle school. Though far beyond the scope of this book, both authors are happily married, and this advice applies here too: Some of the best advice we can give you and your significant other when you're starting out is to put some distance between you and the environment you grew up in. It doesn't have to be drastic, and it certainly doesn't have to be forever or even a long time, but you should learn to see the world in un-tinted lenses.

I (Sean) mention this anecdote because there were clear signs for me that it was time to move on and test the waters with a new employer. In your own career, you may identify deficiencies in your employer such as:

- Limited growth or unclear career path
- Wage stagnation
- Lack of challenging or interesting projects
- Company culture doesn't align with you
- Poor management
- Weak future for the company
- Your interests no longer align to the company

This list can be a multitude of other factors. What is important for you is knowing where you are compared to where you want to be. In previous chapters, we provided a detailed guide for applying to and interviewing for positions so here we won't be focus on the details of changing jobs, but rather how to know it's time for something different. In a lot of cases, you'll just know naturally, as we've pointed out. In others, you'll need some help sorting out what you think, so here we point you back to the need for community in your career. Mentoring is a huge part of professional and personal development, both in the giving (mentor) and the receiving (mentee) side of the equation. We love helping people develop in their careers and themselves, but we weren't always in a position to do that, and we certainly still need a mentor ourselves from time to time. If you think you're starting to be in a position where you need a change, it's a great idea to seek out some guidance of some mentors. These people can be old instructors or university contacts with whom you have a connection, or they can be other professionals in your peer group. In general, we would suggest you gravitate more toward mentors with additional experience and years in the workplace than seeking out only your true peers as mentors. Additionally, it can be extremely beneficial to seek out those outside your organization for the obvious benefit of impartiality. There are certainly exceptions to these suggestions, so feel free to add in the counsel of your trusted peers and coworkers when and where appropriate.

Once you've made sure you know why you're looking to change, something worth noting is certain industries have a much smaller pool of employers to choose from, and thus frequently those industries contain engineers who have bounced back and forth between the three or four available companies depending on who was the highest bidder. While we don't necessarily recommend it as a strategy, I (John) have worked in two industries now where the employees seem to always be on a 3-to-5-year cycle and they just rotate from company A to B to C, often returning where they left in higher positions with larger salaries. The advantage of working in a niche industry is, for the right niches and skills, there is typically a premium on your knowledge and you can be compensated as such. Meanwhile, an advantage for working in an industry as wide as pharmaceuticals, chemicals, and household goods manufacturing is the oftentimes broad availability of employment opportunities.

This brings us back to our list, and the first step in finding a new employer or position: Identifying why you want to leave your current one. If you hate working for a large corporation, what good will it do to move to a competitor of the same size with similar corporate structure? Once you have that in mind you can utilize publicly available information to filter through what you are looking for, regardless of what that is specifically. This is your career, so it's important to pick what resonates with you when you can. If you're a vegan, you likely won't like working for a meat processing plant no matter how good the benefits are. The data you find can include the number of employees, leadership structure, yearly revenue, product lines, or any news stories that shine a positive or negative light on the company. Depending on your circumstance, chances are you have heard through word-of-mouth which companies are considered the best of the best or the worst of the worst. Use this information to your advantage, but remember rumors are just that and may not be indicative of how a company is in all actuality. Again, there are plenty of sources like Glassdoor and Reddit where people air all kinds of dirty laundry from every employer on the planet, but use good judgment and take all of this with a very healthy grain of salt. Usually, there are a disproportionate number of disgruntled people voicing their grievances, so even extremely valid ones can seem more frequent or severe than they really are.

From here you can utilize the job search tools noted in previous chapters to search these companies for available positions that match your skills. You can then apply for the job and start the process, using your experience during the interview process to ask the right questions and ensure the new company fits your needs. Remember, this is an interview for them as much as it is for you, and by nature of having experience and current employment you have an upper hand in the situation that may not have existed earlier in your career. In this case, you very fortunately get to be pretty picky, so we encourage you to take your time and find and apply to roles that actually interest you. Don't be afraid at all to apply for a position above yours if you are due for a promotion, as employers will frequently hire an Engineer I with a few years of experience into an Engineer II role. I (John) have, on multiple times, been hired for roles that were a big jump above my current role, because I was ready to take a leap of faith, and I found places that were willing to take a chance on me too.

An additional step that can be taken before or during the interview process is identifying people with similar positions on LinkedIn that work for the company you are interested in and ask a few questions that will help you make a decision. It is worth stating that no company is perfect and there will be things you love and things that you don't, so think long and hard about why you wish to leave your current position and be sure to ask the right questions to make sure the important things match your desires and goals. Perhaps after listing out all the reasons why you wish to move on from your employer you come to the realization that it is not just your employer that you no longer mesh with, but the industry as a whole. This is a complicated situation for many engineers as the further along in your career you are, the less of a chance you may feel you have to transition industries, which has some truth to it.

It is a bit paradoxical but major industry changes are much simpler for early career engineers as they can take entry-level positions and "start fresh" if you will. That being said, do not let it dissuade you from trying to make a move that you feel best fits your goals and desires. I have spoken with many engineers who feel that after a few years of working in an industry they are tied to it for life. The reality of the situation is as engineers we are trained and knowledgeable in the world of engineering, and just because you worked in manufacturing for 5 years does not mean you have zero transferable skills to move into something like aerospace. Much like finding a job in your industry, you must review the job postings and match your experience and resume as best as possible. When detailed with the correct intent, there are very few positions in engineering that you cannot relate experience to regardless of industry.

Alright, so looking across industries, where do engineering jobs differ and where are they similar? While this answer will be a bit different depending on your subsect of engineering, on a base level being an engineer is the same everywhere. Design engineers are given a problem and must use their technical expertise to create a product or process that accomplishes the goal while maintaining a set budget and timeline. Manufacturing engineers are told to scale a design for production and minimize cost. Program managers and project engineers are expected to orchestrate the entire process from project origination to program closeout. You get our drift; regardless of industry, there are similarities across disciplines.

Among what differs includes what is being worked on (obviously), the tools used to accomplish the project, and the guidelines one must follow when working on a project. The first may be obvious, as emphasized, but switching industries means you are switching product lines, usually in a major way. Sure, one could argue that a machined part is a machined part and the design process doesn't really differ that greatly but being a part of a major project that you identify with deeply can have a massive impact on your mental health when putting in hour after hour on work. An environmentalist at heart will likely wake up enjoying their work day much more at a firm looking to design the next clever invention to clean up the great garbage patch in the Pacific Ocean as compared to designing heavy machinery headed for a fracking site. Passion for your projects can be the difference between going through the motions and thriving in your career.

Another difference between industries is their choice of engineering technology being used. While not a hard and fast rule, there is likely to be commonality across an industry in software. For example, the automotive industry is largely powered by Catia for CAD and may request their potential candidates exhibit experience in this software to be considered for design roles. This can also be seen when moving between companies in the same industry but of vastly different employee numbers. Larger companies typically use programs like Creo and Windchill that can be multi-functional in nature, and cover a lot of individual aspects. You'll also come to find out eventually that the software packages you learned in school, or even what was available to you to use in school, won't be the end all of what you'll need in the work world. Project managers and other project professionals may have been exposed to Microsoft Project in school, and they may have never seen it before going into the workforce, but even with it being the "standard" project management tool, my (John) whole career has been in projects, and only half of my employers have used Microsoft Project in any capacity. Some software packages aren't even available to you outside of the DoD or some government agencies, so expect you'll need to continue learning and evolving your software toolkit as you change companies and industries.

What may be the most underrated difference between industries is the rules and regulations that guide your efforts in a project. Whether it be the FDA, FAA, ATF, DOT, or any other major oversight agency, some industries are held to strict standards with major ramifications for failure to comply with these guidelines. This is a good thing as they are in place to protect not only the manufacturers from liability but the end users of these products, ultimately saving lives and frustrations. That being said, not everyone loves the red tape and hours of specification review required in some industries such as defense and aerospace, so having this in the back of your mind can help guide the direction you wish to proceed in your career.

From here the next obvious difference is the tribal knowledge inherent in more niche industries. For some employers, they have a handful of subject matter experts with intimate knowledge of processes or best practices that will be lost to the ether when these employees either retire or move on from that company. This can result in a sort of stigma that only people with experience in that industry should be hired, which can be a bit difficult for those breaking into the industry as how are you to get experience unless someone gives you a chance to gain that experience. Fortunately, you'll likely find that your second job within an industry was a lot easier to land than your first. That said, we will explain how you can frame experience that may be only tangentially related and make it clear to a prospective employer that you can be an asset to their team.

Fortunately, a lot of things will transfer across industries pretty seamlessly in your career, depending on what knowledge you have exactly, and where you're headed. Some things, like project management, are virtually the same across industries, regardless of whether you're making a loudspeaker or a commercial avionics test solution. Don't believe me? In my (John's) career, I've done just that, and everything will follow a generally similar pattern, where we have some level of development, some level of testing and integration, and general gates and steps to release this product. It's been easily transferred across my career, and it will continue to be transferrable as I progress in my career. If you have a lot of mechanical design experience on the other hand, that knowledge will transfer with you, even if you aren't completely sure of the nuances in what you're developing. Say you are trying to jump industries and you think to yourself

Wait a minute, I spent two, three, four years learning how to design machining fixtures for a job shop. There is no way I will get chosen for a job in aerospace over another candidate who has similar experience but for a company within the industry.

While it may seem that some industries have a major barrier of entry, this is not always the case. Chances are you have many if not all of the qualifications they are looking for, you just have to untangle your experience to best fit their requirements. What does this mean? Well, take the example above. From an outside perspective, machine fixture design is not the same as, say, designing components that are meant to leave earth's orbit. To be completely fair, they are wildly different in use but there can still be many similarities in how the engineering process is applied to the job at hand. Designing machine fixtures will still require the usage of CAD skills, application of GD&T, materials selection, and other related tasks. Just because one is not the verbatim copy of the other, doesn't mean you can't apply the skills. It may not seem like it, but there is a lot more in common with airplanes and submarines than you think there would be. When it comes down to it, both are vessels that move through a fluid.

For this particular example, we know aerospace components are (among many things) typically very specific metals and materials chosen for their mechanical properties, finished in any number of coatings and heat treats, precision built, and abide by a multitude of specifications. Now, as an engineer from a job shop, you can look at all of these components and realize how much interaction exists. While designing build fixtures, you must take into account the base material being machined and adjust your holding force for a jaw, or the material and hardness of the tool so as to not damage or corrode the part, and you must maintain tight tolerances to ensure reliable output of a component that consistently falls within the tolerance of the part itself, while interpreting specifications called out on a customer drawing that will dictate which processes you must follow when working the job – you'll start seeing connections between different worlds. The point here is do not get bogged down in the small details of what you are engineering, but instead deeply examine your process and let the larger picture of being an engineer shape your conversation.

While this example is specifically pointed toward a design engineer looking to move from a smaller industry into one of the more "prestigious," it applies for all subsections of engineering across a very wide variety of industries. As we mentioned, the key is to relate your skills and experience to the engineering process, or even more broadly, the work life of a full-time employee at an office job. Maybe the position is responsible for leading a team but you have never formally been given authority over any other employees, but for an integrated project team you are the go-to person for gathering all of the efforts of the team and providing it to the program manager. With that may come the responsibility of herding some cats, going person to person and checking to see if they are hitting deadlines, and calling meetings to solve problems. All of these examples are informal leadership experience that can be given as proof that you have what it takes to lead a team, even if you have yet to be given the full authority to do so. Perhaps you are looking to move to a company that makes more complex components than your current but you have proven success designing for and using GD&T, which applies to every industry and every manufacturing process.

The point being process, process, process! You may not know how to make a certain elaborate meal but we bet you know how to gather ingredients, wash and cut vegetables, mix spices, and apply some heat. All that is missing is the fine details, the actual process of cooking is something you can have down pat.

Now, like most everything else we talk about in this book, your career mileage may vary and you very well may stumble upon a job that just so happens to be looking for someone with 15 years' experience working with cruise ship boat anchors and no matter what you have done previously, it won't be good enough for the employer. A good thing to note here is that it is not uncommon for a hiring manager to already know exactly who they want to be hired and to shape their requisition to exactly fit the experience of this person. Is it fair? Maybe, maybe not, those people are clearly intended to be hired and a company has the right to be very specific in what they are looking for. These positions are frequently not "real" anyway as they are just a means to an end to get someone very specific hired.

Regardless of whether or not you are attempting to leave an industry or company or stick with what you have, one of the more divisive choices an engineer will make for their career is waiting. This choice being whether to go into management or continue on in a technical role. There is no right answer to this as everyone has different skills and strengths to rely on but there are major differences between the two pathways.

Climbing through the technical ranks is the most straightforward move for an engineer, as you are largely expanding upon your role and responsibilities that you have been working with for your career as an Engineer I/II/III. This career path is ideal for those engineers who love the day-to-day of their job, crunching numbers, and working directly with the products and systems in which they are assigned. Typically, high-level technical roles come with the status (and expectation) of being the mentor and subject matter expert (commonly known as a Subject Matter Expert, or SME in the professional world) of your department. Having spent decades in the technical role you are the go-to person for the hardest challenges and most complex engineering projects. In addition to leading the technical aspects of a program, you will likely be assigned more junior engineers to handle the low-hanging fruit of the project. While the engineers may answer to you on a project level, you are not responsible for the typical managerial responsibilities, and it is very likely that each program has a different mix of junior engineers answering to you, or none at all.

The alternative to this is moving out of a technical role and into an interpersonal one as an engineering manager. Engineering management is an interesting career path as you need skills more closely associated with a business degree while also being able to process the technical aspects of your department. Many engineers find themselves drawn to the managerial route as it can lead to major career strides toward positions like VP of Engineering; however, the truth is many engineers come to miss the actual engineering work over the business-oriented responsibilities. That being said, if you love working with people and helping shape the future of your department, managerial roles are for you.

So which pathway is for you? Technical or Management? In some situations, a person with an affinity for the technical will perform so well that upper management hands them a management position that frankly they are not fit for. While it is important to have skills in both aspects, it is not always possible for a great engineer to be a great manager, and some great managers may not have been the strongest technical engineer. The deciding factor should be determining where your strengths lie. Do you wish to lead teams, hold meetings, budget projects, and use your interpersonal skills to guide your team or would you rather stay hands on, keep up to date with the latest technology and methods used in your industry, and keep your responsibilities held to the engineering level?

Despite the many ways your career can pivot or transition into something new, you may have noticed a trend in our advice that is relevant across any career move. Be aware of the why behind your career choices. There are times when you may be in the middle of your second, third, or fourth year at the same job and start to wonder what else there could be out there for you. This is not a bad thought to have, as complacency is the enemy of growth; however, there should be more than just a feeling or urge for change. Without the correct motivation it is easy to make a change that turns out to be for the worse.

What do you need to identify then to ensure a decision made is the right one? The first thing to do is write out all the reasons you are looking for a change and what has been driving that feeling. For example, your current department is dysfunctional and you want to get out before the wheels fall off. What is driving that feeling of dysfunction? Are all the programs in disarray and way behind schedule? Does your manager make awful decisions at every turn? Does the company as a whole have a poor outlook in the next few years? Or perhaps you are experiencing a rough patch in a project you are working on and commiserating with coworkers has led you to believe everything is much worse than it truly is? There can be a level of negativity that is perpetuated by employees when they have issues arise or have to do extra work because someone else made an error. For some, this is just a way to blow off steam and share in their annoyance with the current situation but for early career engineers this kind of conversation might result in anxiety over their jobs which then snowballs into thinking that you have to leave before something even worse happens and you are left in a failing company or get worked to death trying to fix the ship before it sinks. That is why it is important to know what is behind these desires for change, some problems may seem overwhelming but may actually be short-lived and easily explainable.

Once you have a well-thought-out list with all the motivations to leave you can counter this list with a list of expectations and hopes for the next phase of your career. A pro and con list of sorts, where one side describes all the things you wish to avoid and the other all of the things you wish to have. Visualizing what truly makes you fulfilled and what you dislike most about your career is a powerful decision-making tool that will give you the clearest path forward.

As stated earlier, major career transitions are all but guaranteed in this day and age so arm yourself with all the knowledge you can to make the best choices. The upside to moving jobs and positions within the engineering world is that you are constantly being challenged to find new answers to problems you may have never been faced with had you stayed complacent. As engineers, there should be an innate desire to seek out the unknown and make it known. You may find some of the greatest growth in your professional and personal lives within the discomfort of transitions. Regardless of whether or not you are attempting to leave an industry or company or stick with what you have, one of the more divisive choices an engineer will make for their career is waiting for you to address it. This choice being whether to go into management or continue on in a technical role (again, these are not the only two possible outcomes in your career, but this question comes up often enough we wanted to address it here). There is not only one right answer to this question as everyone has different skills and strengths they bring to the table, but there are major differences between the two pathways.

Climbing through the technical ranks is the most straightforward move for an engineer, as you are largely expanding upon your role and responsibilities that you have been working with for your career as an Engineer I/II/III. This career path is ideal for those engineers who love the day-to-day of their job, crunching numbers, and working directly with the products and systems in which they are assigned. Typically, high-level technical roles come with the status (and expectation) of being the mentor and subject matter expert (commonly known as a Subject Matter Expert, or SME in the professional world) of your department. Having spent decades in the technical role you are the go-to person for the hardest challenges and most complex engineering projects. In addition to leading the technical aspects of a program, you will likely be assigned more junior engineers to handle the low-hanging fruit of the project. While the engineers may answer to you on a project level, you are not responsible for the typical managerial responsibilities and it is very likely that each program has a different mix of junior engineers answering to you, or none at all.

The alternative to this is moving out of a technical role and into an interpersonal one as an engineering manager. Engineering management is an interesting career path as you need skills more closely associated with a business degree while also being able to process the technical aspects of your department. Many engineers find themselves drawn to the managerial route as it can lead to major career strides toward positions like VP of Engineering; however, the truth is many engineers come to miss the actual engineering work over the business-oriented responsibilities. That being said, if you love working with people and helping shape the future of your department, managerial roles are for you.

So which pathway is for you? Technical or Management? In some situations, a person with an affinity for the technical will perform so well that upper management hands them a management position that frankly they are not fit for. While it is important to have skills in both aspects, it is not always possible for a great engineer to be a great manager, and some great managers may not have been the strongest technical engineer. The deciding factor should be determining where your strengths lie. Do you wish to lead teams, hold meetings, budget projects, and use your interpersonal skills to guide your team or would you rather stay hands on, keep up to date with the latest technology and methods used in your industry, and keep your responsibilities held to the engineering level?

Despite the many ways your career can pivot or transition into something new, you may have noticed a trend in our advice that is relevant across any career move: **be aware of the why behind your career choices**. There are times when you may be in the middle of your second, third, or fourth year at the same job and start to wonder what else there could be out there for you. This is not a bad thought to have as complacency is the enemy of growth; however, there should be more than just a feeling or urge for change. Without the correct motivation, it is easy to make a change that turns out to be for the worse.

What do you need to identify then to ensure a decision made is the right one? The first thing to do is write out all the reasons you are looking for a change and what has been driving that feeling. For example, your current department is dysfunctional and you want to get out before the wheels fall off. What is driving that feeling of dysfunction? Are all the programs in disarray and way behind schedule? Does your manager make awful decisions at every turn? Does the company as a whole have a poor outlook in the next few years? Or perhaps you are experiencing a rough patch in a project you are working on and commiserating with coworkers has led you to believe everything is much worse than it truly is? There can be a level of negativity that is perpetuated by employees when they have issues arise or have to do extra work because someone else made an error. For some this is just a way to blow off steam and share in their annoyance with the current situation but for early career engineers this kind of conversation might result in anxiety over their jobs which then snowballs into thinking that you have to leave before something even worse happens and you are left in a failing company or get worked to death trying to fix the ship before it sinks. That is why it is important to know what is behind these desires for change, some problems may seem overwhelming but may actually be short-lived and easily explainable.

Once you have a well-thought-out list with all the motivations to leave you can counter this list with a list of expectations and hopes for the next phase of your career. A pro and con list of sorts, where one side describes all the things you wish to avoid and the other all of the things you wish to have. Visualizing what truly makes you fulfilled and what you dislike most about your career is a powerful decision-making tool that will give you the clearest path forward.

As stated earlier, major career transitions are all but guaranteed in this day and age, so arm yourself with all the knowledge you can to make the best choices. The upside to moving jobs and positions within the engineering world is that you are constantly being challenged to find new answers to problems you may have never been faced with had you stayed complacent. As engineers there should be an innate desire to seek out the unknown and make it known. You may find some of the greatest growth in your professional and personal lives within the discomfort of transitions. Ultimately, transitions are just one more fork in the road on the journey of your career, and they are not something to be feared. After all, this is your career, and we hope you make the best of it.

CHAPTER SUMMARY

Career transitions are anything from minor changes, like being promoted along your career path, to major changes like, switching industries or roles entirely. Major changes have become the norm, and it is important to make these moves in the most educated way possible. To do this, identify what is driving you to make these changes and utilize that information to guide you from this phase to the next.

- Create a list as a decision-making tool that will aid in visualizing what truly makes you fulfilled and what you dislike most about your career
- Major industry changes are simpler for early career engineers, but it is not impossible for experienced engineers to transition to other industries and relate their skills and experiences
- Mentors are integral to professional and personal development; they can be old instructors, university contacts, or other professionals in your peer group
- The upsides of moving jobs and positions include being challenged to find new answers to problems that you may have never been faced with before
- Changes fight complacency and help ensure you're always growing and developing
- Making a major career move requires identifying what is driving your desire for change and counterbalancing it with a list of expectations and hopes for the next phase of your career
- Be aware of the why behind your choices

10 Things We Wish We Knew Sooner

HOW DO I APPLY TO A SCHOOL?

In some cases, you've always known you were going to college, and in others, that wasn't the plan at the beginning. Over the course of our mentoring, we've run into what is basically the gamut of students, from the very traditional to the very non-traditional paths on can take in going to college. Furthermore, we talk to a lot of engineers and students who are the first in their families to go to college, so we fully understand this path isn't always clear. You may be really on your own, and wondering how you even apply to a school in the first place. Fret not, because we have you covered. Once you know where you want to go (or at a minimum, a short list of where you want to go), you're going to have to apply. We recommend only applying to programs you're actually interested because (1) college applications are a lot of work, and (2) they aren't free (usually). If you're applying to school after school, those fifty and sixty-dollar application fees add up quickly, though we do want to point out there are a number of ways to apply for free, or at reduced costs, like being military, a first responder, or the child of an alumnus in some cases. Take a look and see what each school offers when you're looking.

When you have a list of those schools, you'll need to gather your supporting documentation. This is usually your transcripts (high school and junior college if applicable for undergrad programs, undergraduate transcripts for grad school), a well-written statement of purpose, any relevant entrance exams (like the ACT or SAT, or GRE for grad school), and get a few people to write you letters of recommendation. For the statement of purpose, you're going to need to clearly and concisely state why you chose to pursue this specific educational path, and what it means to you. This is an excellent way to allow your passion to shine through, so don't be intimidated by writing one! As for your letters of reference, usually, it will be from people who can speak to your work and capabilities, so think along the lines of past teachers, instructors, and bosses/managers, both past and present. You'll usually be barred from asking friends or family from submitting letters on your behalf, due to potential biases. Make sure you follow some general rules of reference etiquette, such as giving your references plenty of time to get one done (think a couple weeks or so), as well as a gracious way to bow out if they decline to be a reference. There is always a possibility that they don't think they are as good of a judge of your performance as you think they may. Not all schools require statements of purpose/essays/reference letters/entrance exams, and not all programs in those schools require the same things, so check with each department. In our experience, there's usually some really helpful resources at either the school or department level who can help guide you, as this is what they do as their regular job. Don't be afraid to ask questions, or be intimidated by the process. It's a

 DOI: 10.1201/9781003510901-10

lot of work to pull together, and it's going to take you some time to submit an application, but once you're accepted, your academic journey can begin just like that! In the event you don't get accepted into the program you want, we know that can be disappointing, but don't let it get you down. There is a lot of subjective reasoning involved in the decision process, and it's not necessarily a reflection on you per se. I (John) have been rejected from programs before, and other (better) doors opened up later on. That's just part of growing. Once you get accepted, there are all kinds of additional things you'll need to do before you begin school, but your institution will tell you specifically what you need to do in those cases, and you may even have a person assigned to you as you transition in.

WHAT IF I'M BAD AT MATH? CAN I STILL BE AN ENGINEER?

This is an interesting one, and we used to hear friends of ours say various forms of "I could never be an engineer because I'm bad at math." Interesting because the answer may surprise you; it's not an outright no, it's a solid "it depends." Let me explain. I (John) have known plenty of engineers and students who just think in a mathematically focused way. They never struggled with math, or seemed to have any difficulty grasping concepts. Even well into calculus and differential equations, they just got it. I am not one of those people. I struggled with algebra to some extent, which made calculus difficult. In fact, when I struggled with calculus, it was more often than not the algebra that I struggled with instead of the calculus-specific portion. Statistics was not intuitive to me in any shape of the question, but thankfully, I always understood trigonometry really well. Trig ended up being one of my saving graces during my undergrad, because I could frequently break things into sines and cosines when I got stuck. I also, for better or worse, am one of the most stubborn people I know, and that made me unwilling to give up. I remember being in my undergrad and saying I would leave that town either dead or as an engineer. A bit dramatic to be sure, but I said I am stubborn, and that allowed me to bludgeon my way through my required classes and squeak by. Squeaking by on account of stubbornly bludgeoning my way made things a lot harder than it had to be, and that leads me to my point: Not everyone gets math, but everyone can get better at math. I made things way harder on myself because for a long time, I wasn't willing to put in the work and effort to make it happen. The results of that resistance to effort, we've already discussed – I got kicked out of a program. Math may not come easily to you, but the correct response is not to just give up on things that don't make sense to you.

You'll have to decide how much you really want to do this because there is a lot of math involved. A lot of generally hard math, actually. You're going to take at least three semesters of calculus, plus probably things like statistics, differential equations, linear algebra, and maybe even some finance classes. Being good at math is a huge benefit, but being willing to learn and overcome adversity is a bigger benefit to you. If you struggle with math, you're going to feel like you're beating your head against a wall some days, and you'll just have to suffer through some classes that don't click. The good news for you is there are multitudes of resources which are available now, that may or may not have been available when we were doing our undergrads. From on-campus tutors to literally dozens of YouTube channels where people explain step by step how to do everything from basic algebra to masters-and-PhD level math, you can get help. Your first step is to admit you need help, then not be too proud to ask for help. Your pride doesn't help you at all here, and in fact can only hurt you because so many things build on these foundational math classes. Once you decide your desire to earn a degree and your stubborn resolve are stronger than the math you're facing, there's a huge silver lining for you to look forward to: You'll very rarely do the math you just don't get out in the real world. I have used calculus only a couple times since graduating, and most of my math is now applied concepts and analysis. Certainly, this will depend on what you're doing, but it is absolutely possible to be an engineer and not do the math you hated in college, or be an engineer if you don't love math. Most of what I'd guess you struggle with is the pure math, and not the applications, so hold on. Once you get to your major classes, you're likely to see how to use it in the applications of what you're interested in, and not just in raw form.

In short, you absolutely can be an engineer and not be spectacular at math, but it's not possible to be an engineer who didn't put forth the effort to get through those math classes.

I FAILED A CLASS, (OR WORSE) – WHAT HAPPENS NOW?

If this happens to you, you're probably going to be devastated. Or if you're like me (John), then you're going to go down a dangerous rabbit hole of self-deprecation and tell yourself that you're a failure. I told you I was highly dramatic. Nevertheless, I have failed more than one class in my life, and as much as it makes you doubt your own ability, it's not the end of the world (I've actually had to retake several classes several times each. I'm not proud of that, but I can safely say that it's really not the end of the world.) There are certainly consequences though, and you'll have to deal with them. They can take many forms, like hurting your GPA, delaying graduation, costing you more money, or just hurting your own self-esteem, to name a few.

If you fail a class, first of all, give yourself some grace and some time to come to terms with that fact. Next, identify what happened to make this the case, so you can fix it in the future. Were you just slammed because you took 21 hours of credit and this one slipped through the cracks, or did you just not get the material and you were either too embarrassed to ask for help or just too lazy to put in the work? Those are all vastly different circumstances that lead to the same outcome. You're going to need to identify which one(s) got you in this predicament, and correct it the next time. The best-case scenario in failing a class is you just get back on that horse next semester and do fine, and everyone goes on their merry way. Depending on your institution and circumstances, you may even be able to do what's called a grade replacement, and once you've passed, you can swap your D or F out for hopefully and A or B, and that's that. No real harm is done, you just had to pay for a class again. That doesn't always happen, and sometimes you can throw off your graduation if you have to retake a class that is only offered in a specific semester or term. Grade replacements are also usually limited to only somewhere around 15 hours of credit replacement, or about five classes, so use this plan sparingly. The actual worst case is you do as I did and fail enough classes (even including grade replacement) that you get kicked out of a program. We hope you didn't experience the full worst case, but if you did, it's more or less fixable too, you're just going to have to work like crazy to dig yourself out of that hole. In all likelihood, you're probably somewhere in the middle, and you didn't get away basically scot-free, but you didn't get kicked out of a program either. If you find yourself having failed a class, you're not alone, even if it feels like it. There are a surprising number of your fellow engineers who failed at least one class, and, for some reason, it's virtually never discussed in engineering circles as some taboo. If you made it through engineering school without ever failing a class, that's awesome. If you did fail a class, you're still in this, and you can have a very successful career, both in and out of school. If you failed a lot of classes, you can also still be successful, but you have to make changes and take this seriously. Despite how it feels at the time, it's not going to be the end of the world, but like most learning and growing opportunities, it's going to hurt a bit, or at the minimum be uncomfortable. If you're in this situation, without question, the best thing you can do for yourself is to identify what led to you failing a class and fix that to the best of your abilities before trying again. Fix what you need before you get kicked out of a program, and this simply cannot be overstated. You HAVE to get things squared away in any way possible. If you can correct the circumstances that got you there, you can move forward again. As long as you make it through, you'll look back on those dark spots in your memory as just little blips on the radar.

I'M WORRIED I PICKED THE WRONG MAJOR, AND I DON'T KNOW HOW TO FIX IT

A lot of these situations are nuanced, and as such they don't all have super straightforward answers. What makes you feel as though you've selected the wrong major? Did you get into chemical engineering and realize you hate chemistry, or get into chemical engineering and realize you wanted to be a children's book author instead? It's important to know why you feel it was the wrong one, so you can do something about it. We understand work is a huge part of being a human, so we don't want or expect you to do 40 years of something that makes you miserable. There are a few ways we could see this going, and we'll try to cover each one.

If you just decide you don't want to be an engineer, there's going to be remarkably little we can do to convince you otherwise. If it doesn't click, doesn't spark your interest, or if it's just not worth the effort, then we'd likely think your heart wasn't in it from the beginning – that you were trying to pursue someone else's dream for you. No amount of persuasion will make you want to do something your heart's not in. In that case, it's going to be better to jump ship sooner than later, ideally before you've gone too far down an academic path. Fortunately, the earlier you are in college, the easier it will be to use those classes without having "wasted" them (a mechanical design class won't help you graduate with an arts degree, except as an elective, and you may have used all your electives by then). In this case, there's no harm and no foul if you and engineering part ways. It's not for everyone, and that's okay. The world needs all kinds of passionate people, so we want you do be where your passions are.

If you are still interested in being an engineer, but you just don't like the major you picked, that's pretty easy to fix too. It's even quite common, as nobody really knows what they like until they start doing it, so don't feel too bad if what you graduate with isn't necessarily what you started with. Even I (John) did that, and finally decided to pursue my original target a full decade after changing majors. Depending on where you are in your college career, a lot of the classes will transfer or satisfy requirements, so you may not even have much of a bump in your path. Again, the earlier you can identify what you don't like, the easier it will be to fix. What would be the hardest for a student to fix would be to be in your final semester and decide you didn't want to finish your major, and here we'd caution you quite a bit. If you're less than a semester away from finishing your degree, it's more than likely going to be better to just finish it out and adjust from there. If you wanted to go to something different entirely, then you're going to need to understand the cost associated with that change, like additional classes, delayed graduation, a longer time in whatever housing situation you're in, as well as the potential for more student debt, which we want to help you avoid.

There is so much to consider before you decide to just quit so close to the finish line, so we want to make sure you know why you want to quit. It may not be a fun thing to hear, but having a degree you don't like is far better (read: more useful) than having 90% of a degree you don't like and starting over. You can always pursue what you like on the side, or while working full time. In general, being a graduate of something you don't love still leaves you with a lot more in the way of options than not being a graduate of something you don't love. In our mind, this is going to be a question of how much do you have into it at this point (time, money, etc.) versus the alternative. There's a time to not play the argument of sunk costs, but to us, this is not it. The reasons we say to not change your major too late in the game is because you lose opportunity and that can compound as you go. What do we mean by that? Well, if you don't graduate, you're not going to be able to land a job that has having a degree as a true minimum requirement. That means at the least, you're going to lose out on income potential, and you may then have to extend your time in college, take on (more) loans, or something similar. What if you could work and use some level of your experience to get some income going, and then work on how to change what you don't like? If you can earn money, you have a lot more wiggle room available to you to change what you don't like.

When working through things like this, the sooner you can get a handle on it, the better off you are, so again, we encourage you to be proactive and really consider what you're wanting from your career. You can fix anything, but it can be a lot more work after a certain point.

HOW CAN I GET MORE HELP?

You're going to need help at some point. You will. There is no way around that. You will need it, but you won't always know how to go about getting it (or honestly, even when you'll need it). We're going to look at a few of the most likely situations where you may need help, and if you're in a situation that doesn't fit in one of these exactly, the methods to get help are generally the same, or at least similar.

1. Getting help in/with school
2. With friends/family who don't understand

3. At work
4. Maintaining a work–life balance

We'll tackle these in order, so to begin, how do you get help in school? This is probably the first place you'll need help, not only from the obvious reason that engineering school is hard but also because this is likely the first time you've been somewhat on your own/in a different environment than your high school peers/ actually needed help, or any combination of the above. Maybe you're the first person in your family to go to college, or the first to try engineering school, or maybe English is your second (or third and beyond) language, or you're even new to American culture. This of course assumes you're going to college in the US, but even if you're not, these approaches will still apply. Also, if any of these categories do apply to you, we applaud you for getting to this far and taking some chances. There are a few things you could need help with in college, but we'd assume most of them fall under either needing academic help, adjusting to the college environment (i.e., being more or less on your own for probably the first time in your life), or needing help with managing your time. The good news is you're going to be surrounded by peers who need similar help, and most schools offer some level of assistance in this regard.

If you need help with the academics, admit it. There is absolutely zero shame in saying you need help understanding something, and even less reward in waiting too long to get help, trust us (as evidenced from the anecdote at the beginning of the book, I (John) am speaking from real experience here.) The earlier you can admit it the better, and you'll appreciate leaving yourself with a lot more runway left than the end-of-semester approach we've all been guilty of where you do some finger-counting math and say "if I can make a 120 on this next test, I can still pass with an A." Once you've admitted you need help, you can then look to see what is available to you, either from the school, your instructor, the internet, or your peers. Remember, if you don't understand something, it's entirely probable that someone else doesn't understand it either. Depending on the subject, size of your school and some other factors, your school may even have structured tutoring on those classes, and you may not even know it exists, so we encourage you to look into it and ask your instructors. Those sessions were called "LEAD" sessions in my (John's) undergrad, and to be honest, I never went much because they were led by students who were generally oddly smug about knowing better than we unenlightened, so I get it if that's not your first choice. What do you do in that case? You're going to need to find an alternate source, but, fortunately, the internet is a lot more useful now that it even was the decade before, when we were in your shoes, so you can find tutoring of all kinds if you're willing to look. If you need help, you may also just need to do a lot more homework than is assigned, so an option here is for you to recruit your friends and classmates to work with you and all study more than the minimum. Truly, at this point, admitting you need help is the first step, and form there you can identify what works for you and develop a plan to attack it.

If what you need help with is time or time management, then as we've said before, the most important thing you can do in this case is develop a strong routine you can stick to. The specific *what* of this routine doesn't matter nearly as much as the *how* it works for you, so consider what you need from a routine and plan accordingly. What this means is if you want to work out for example, but you absolutely hate getting up in the morning, then maybe planning your day around a morning workout is dooming you to fail. Could you schedule in your workout in the middle of the day, the evening, or even at night? We'll hit this in detail in the work–life balance portion, in just a few paragraphs.

What about when what you need help with is just feeling like people don't understand? Your friends at other schools and in other programs always seem to have fun on the weekend while you're stuck in the lab, or your parents don't exactly understand how you managed to fail a class? This is pretty common to some extent, so don't be surprised to run into this eventually, at least once. As you probably remember from earlier in the book, I (John) failed some classes. Actually, a bunch of classes. I had extremely supportive parents, and I knew they would be behind me 100% of the time until I finished, and that was amazing. I also know that not everyone has that kind of support system from family, or from friends, and that can make things a little more interesting for you. If you're not feeling supported by your family, this can have some major impacts on you directly, from financially supporting you to just feeling like you have to do everything on your own, but if you're not feeling like you have the support of your friends, you may just feel like the odd one out. What do you do? If you're a person whose support system hasn't gone through a challenging major like engineering (or possibly much less school at all), then it's going to be extremely hard to get them to understand what you're going through. It's more often than not something that a person has to experience for themselves to really understand, so don't think that you're going to just make things click for them in a conversation or two. You can absolutely talk to them, explain why things are the way they are, even show them what you're working on if they are interested, but there is still a really good chance that they'll never fully get it, and that's okay. Your family and real friends will support you in what you're doing, even if they don't always understand your why. If you're in a situation where you're financially supported by your parents, though, and they are about to cut you off because you keep failing classes, that's a different story entirely. Hopefully, you're not in this place, and if you started down that path, you were able to recover to a point where it didn't happen, but we understand that hope is not a strategy, and boats don't float on hope. If you find yourself in this spot, then chalk that up to the category of things you didn't think would happen which we mentioned in a prior chapter. All actions have consequences, and not all consequences are good ones. Again, we'd ask you to take an internal look and see how you ended up here, and what you can do to fix it, as well as ask yourself how badly you want to succeed because it's going to be harder and harder the less support you have. Fortunately, that is kind of an extreme case, and in our experiences, you're just more likely to have friends who don't understand why you're always busy. In that case, maybe they will come around, and maybe they won't, but the reality is as long as you understand your why, then you have enough to succeed.

Ok, great. You've made it through school and family and friends all of that, but now you're out of school and you need help at your job – the kind of help you're embarrassed to need. Kind of like when you've met someone more than once or twice so you feel like you should know their name by now but you don't, and now it's way too late to ask. The uncomfortable truth here is that if you're in this situation, you're going to need to get yourself squared away sooner than later, so you need to acknowledge this one fast. If it's something work-related that you feel you can get help with at your employer, do it. Bring it up to your management as something you need help with, because part of their job is to make sure you have what you need to be successful. If it's something that you don't feel like you can bring up with your management, then you're going to need to take it upon yourself to catch up on outside of work, either by practice, formal, or informal education, or a combination thereof, and the faster you can do this, the better. In this case, whatever your shortcoming is won't just magically fix itself, and the sooner you can right the ship, the better. You may have to take a (potentially) embarrassing short-term loss to make a long-term gain, and those are better to just get out of the way as soon as you can. Don't try to hide that you need help, and seek out people whom you know can help you, whether your management chain, your peers, or something else, don't pretend that you're a-okay and nothing is wrong. You'll thank yourself later for fixing this now.

One of the hardest things we see come up over and over for people is the topic of work–life balance, whether in school or out in the workforce, or both. Time is one of the only resources you can't earn more of, so it makes sense to be protective of it and have all these things you want to do, but then next thing you know, your free time has disappeared, and the day is over. Not only is the day over, but you've done nothing that you wanted to do, and only some of what you needed to do. Sound familiar? This happens a lot, from when you go to school and are more or less alone and left to your own devices for the first time, or when you transition from the no-class-in-the-morning academic life and suddenly find yourself working mornings and 40-hour work weeks. Between school/work, working out, sleep, and a social (media) life, where do you fit everything in? The truth is time management is more of a learned skill than it is an inherent one, and it's something you need to develop. There is the saying we've all heard that "everyone has the same 24 hours in a day," but we don't necessarily agree to this saying the way it's usually presented. Everyone has 24 hours, but they don't all have the same 24 hours. From work to taking care of other responsibilities, whether we chose them or we didn't, we don't all have the same hours to use for the same things. What we do have though, is the ability to make and develop our routines. You may have noticed a recurring theme throughout this book where we encourage you to get in a routine for a lot of things, and this is no different. Regardless of what kind of responsibilities you have, you can set a routine for yourself, even if it's one that's shorter and simpler than you may want it to be, and when you're starting out, that's okay. People often overestimate what they can do with a short amount of time, and underestimate what they can accomplish with a longer time, and that holds true for this as well. Whether you discovered that you need to better manage time from your school experiences or from the workforce, it's a skill that will help you find more enjoyment in your life because it won't feel like you're always crunched for time.

HOW DID WE PICK OUR SCHOOL(S)?

Obviously, to be an engineer, you're going to need to graduate from engineering school. If you're reading this, then you've likely already picked your school, but maybe you haven't, or maybe you know there will be another round of school in your future, so you still have a decision to make. How did we pick our own schools? Does it even matter?

The first thing we'd say to this question is as long as you're going to a real, accredited school and not some kind of degree mill, then your education will likely be adequate. We say adequate because that's the entire point of an accreditation board. Notice we don't say "equal" here, because not all education is created equally. Even within accredited programs, there are varying degrees of quality and perceived quality from school to school, or even within different programs at the same school. Some schools have great civil engineering programs while not having remarkably good electrical engineering programs, as example. Some schools have excellent engineering programs and are objectively terrible everything else, or are great schools for pretty much everything but engineering. The short answer is yes, it does matter where you go to school, but maybe not to the extent you think it does. How do you pick a school then? We'll answer this in no particular order, because you may value one criterion more than we do, or vice versa.

As we just said, one of the first things to consider is whether or not this school is accredited. There are non-accredited schools out there, and some of them may even be good choices, but why take the risk? In the US, ABET (which stands for "Accreditation Board for Engineering and Technology, and it was originally founded by some of the older professional engineering societies in the United States) accreditation is a pretty common one to look for with engineering, but there are other accreditations that have similar requirements and quality, so we don't direct you to look only at ABET-accredited schools.

From there, the other likely factors you're going to need to consider include things like distance or convenience, likelihood of being accepted (either from your previous grades or your test scores – in truth a lot of factors impact this), faculty/student ratios, quality of facilities, overall cost of attendance (school costs more than just tuition!), and even the faculty itself may factor in to your decision as you're picking a school and program. In fact, even tradition can be something you consider – I (John) am a second-generation Auburn Tiger. As we said, we can't tell you which of these is going to carry the highest weight in your decision, so we want to make sure you know enough to make an informed decision. To start, know which of these items has the highest weighting to you, and also know that rating won't be the same for everyone. Once you have an idea of the relative weights for these, it's time for you to do some research: Do you know what technical area you want to be in? Do you know where you want to live geographically? You may not think it, but some schools do have better presence in certain locations/regions than do others, so while not necessary, if you're a long-term planner, this can be helpful to have figured out early. Ultimately, we chose our schools because of which ones scored the highest when weighted against our own criteria, and don't regret our choices. Put in the work you need to thoroughly research and vet out whichever schools you're interested, and you'll make an informed decision too.

I HAVE BEEN AN ENGINEER FOR A FEW MONTHS, HOW DO I KNOW IF I AM DOING ENOUGH?

One of the most intimidating things for me (Sean) when I first started on my path in engineering was trying to make decisions and efforts toward a project without wasting anyone's time or money. Your impact differs strongly depending on how large of a firm you are working with as well as the trust they put in you right off the bat. My first job was at a late-stage startup with two other mechanical engineers and with that came the knowledge that a large enough error could cause serious harm to the up-and-coming company. This made me hesitant to trust my decisions and design choices on even more than the usual uncertainty that comes with being new at something. So, with that extra pressure of indecision, I started to wonder if I was even providing any value through what actions I did take.

While it is true that a company hires you because they expect you to provide a greater value than your employment will cost, it is also true that any good employer knows there is a level of ramp-up required to bring a budding engineer into the fold. The fact of the matter is college prepares you a lot for how to solve problems and analytical thinking, but only experience through solving real engineering problems will allow you to excel as an engineer. My, and maybe your, fears of inadequacy or the feeling that you are not providing any value should be taken with a grain of salt, especially in the first few years of your career.

If you are starting to feel like you are not providing enough value, speak with your manager or supervisor. Ask them what you can do to aid the team or where you can find resources to increase your value. If you are early enough in your journey, chances are they will tell you to shadow another engineer until you learn the processes and methodologies of the department, or maybe you will have to sit back and just observe until a low-impact, low-difficulty assignment arises in the program. Regardless, the best way to change this mindset and expedite your productivity is to follow our recommendations in the Maximizing Growth chapter and dive head first into taking control of your destiny. Engineers are not born, they are made, and with a little time and a lot of effort you will be another highly motivated and productive member of whatever engineering team you choose.

SHOULD I LET AI WRITE MY RESUME OR COVER LETTER?

One of the most tedious and challenging aspects of the job hunt is authoring and personalizing your resume to best align your skills with a job posting. (You'll notice that we wrote chapters about that exact process in this book). While we did our best to streamline the process for you, you may still feel a bit intimidated and are looking for anything to make it easier. In comes AI.

As of the year of authoring this book (2023), artificial intelligence is rapidly growing and improving with little slowdown in sight. People can write code, build websites, converse, and create text or images as detailed as their prompts. While this is extremely powerful, it is also not perfected. AI text generation is largely lacking in originality and personality and after exposure to a few paragraphs an experienced eye can spot the signs. Additionally, many if not most companies have added scanners to their application portals intending to catch AI submissions to clear those out of their queue. Employers want people who do their own work and put in a bit of effort to join their team.

This is to say, don't copy paste a block of text from ChatGPT or whatever your favorite text generator is. Instead, if you are having trouble finding your way through a cover letter or resume, utilize the software to produce an example and review it for its best aspects and put that in to your own words. Add some humanity to it because once you get to a recruiter or hiring manager you really have only one shot to portray your experience in a way that entices that first interview and a robotic copy–paste description probably won't cut it. AI is a great tool, but it is not a replacement for human interaction.

WHAT IF I AM NOT A TRADITIONAL STUDENT?

To this point, we've made a lot of implied assumptions that if you're reading this, you're a traditional student – that is, that you're in college immediately after high school (in the USA). Plenty of engineers and professionals make this decision a little later on (we've been in school with them!), so don't feel like you've missed the boat, so to speak. If you're a non-traditional student, this is great! We're thrilled you joined us, even if you made some other stops along the way. Most of what we outlined in this book will still apply to you, but there may be some changes you'll need to make. For example, if you're in college later on, it's likely going to be a different experience than you'd have had at 20, and that's completely understandable. It may be hard for you to take advantage of some of the avenues and resources we laid out, so if you find yourself in this place, you need to make sure you have a firm handle on skills like time management, identifying what is important, and following through with a plan. Being a little bit later to the party does have some benefits though, and we think you'll agree that a lot of this is easier once you've had the opportunity to learn more about yourself. In short, take what we've laid out here, identify what you need to modify, and then apply it to your situation. As we've said before, engineering is for everyone who is qualified, so if you're hitting this point a little later on, there's still plenty of opportunity for you to succeed. Honestly, once you're in the professional world, you'll barely even notice you joined a little later than the rest of us.

Index

Printed in the United States
by Baker & Taylor Publisher Services